Cobra in the Clouds

by
JOHN STANAWAY

COMBAT HISTORY OF THE
39th FIGHTER SQUADRON
1940-1980

This is not a reprint

P. O. BOX 33
TEMPLE CITY, CALIF. 91780

ISBN-0-911852-92-1

INTRODUCTION

". . .by the fourth day, December 13 (1941), we dismissed the possibility of any form of enemy counterattack. In three days our Zero fighters had given us absolute aerial superiority in this theater of war."*

Thus two Japanese leaders of their country's effort in World War !! accurately described the situation that was to prevail in the Pacific for many months. An inadequate Allied response to Japan's initiative allowed the Philippines and the Dutch East Indies as well as vast portions of New Guinea and other areas north of Australia to fall into Japanese hands. American air commanders, especially, were astounded by the flashing performances of the Zero fighter and were confounded to find effective countermeasures with existing equipment. Within the first four months of the Pacific war no fewer than fifty-eight U.S. Army fighters were lost from a strength of about one-hundred-and-fifty aircraft.

Into this bleak situation the U.S. 39th Fighter Squadron received its baptism of fire. The young pilots of the squadron countered some of the most formidable opposition of the war. With an eagerness and determination that would mark the squadron throughout its history the 39th pilots fought and learned their grim trade. In the process those pilots garnered some impressive claims to fame for their squadron and hammered out an honorable tradition. At least ten victories were chalked up by the 39th in Bell P-39 Airacobras which aircraft were not highly regarded by many pilots. And these victories were won without a single 39th pilot falling in combat. Yes, several pilots were shot down but not a single man lost his life in battle with the Bell fighter. This fact is even more remarkable in light of average casualty rate among other veteran P-39 units of between nine and twelve killed.

Other notable achievements of the 39th mark it as a key unit of the Southwest Pacific. It introduced the P-38 to the New Guinea front and initiated the phenomenal record of that great aircraft. The 39th was the first Fifth Air Force fighter squadron to claim one-hundred or more victories in aerial combat. At the time of the Rabaul campaign in Autumn 1943, the 39th had more aces on its active roster than any other Fifth Air Force squadron. The 39th is generally given credit for the 2,500th V Fighter Command aerial victory and a 39th formation led the first Fifth Air Force sweep over Japan.

**ZERO, by Masatake Okumiya and Jiro Horikoshi with Martin Caidin E.P. Dutton and Co.*

Another unique fact about the 39th is that the top-scoring U.S. aces of World War II and Korea both served in the squadron. Dick Bong scored his first five victories and Joe McConnell all sixteen of his with the 39th. As it did in New Guinea the 39th did the lion's share of scoring aerial victories during one period of the Korean war.

Throughout its career the 39th was also distinguished by the variety of fighter types that it used. It started out using P-35s when the squadron was formed in February 1940. Later it switched to P-39s and the P-400 deriavative which the squadron took into combat. Then came the P-38 and the 39th's great period of fame. Later the squadron did faithful work with P-47 and P-51 aircraft. After the war the 39th soldiered on with the P-51 and entered the Korean War with the venerable Mustang. After a brief tour with a subsequent replacement, the F-80, the 39th went back to the Mustang before re-equipping with a brand-new ace-maker, the F-86 Sabrejet. Since the Korean war the 39th has been operational with a number of different types including the F-94, RB-66, F-105G, and F-4 Phantom.

In effect, the 39th was a special sort of fighter squadron with many varied distinctions. The "Cobra" of the title refers to the insigne of the squadron which was created, reportedly, by the Walt Disney studios during the 39th's tenure with the Airacobra in the U.S. . When the squadron shifted from the Bell fighter the cobra continued to signify the unit itself and its deadliness in battle. From the struggle to protect Port Moresby to the whistling cold battles high over Korea's Yalu river the 39th maintained its honor.

It has been the pleasant experience of the author to interview a number of the 39th veterans. One striking feature of all these men is their pride and grace in reflecting on the 39th. One could expect that either the bravado of passing years or reluctance to remember such a traumatic period would prevail. But the case was that each veteran was eager to help present an accurate account (indeed, Col. Charles King honored my mailbox with weekly packets of diaries, records, personal accounts and anything else that could have been of possible use) and I found that the pilots and crews of the 39th were probably more interested in the humorous activities of their comrades than in the heady battle honors that were most justly theirs.

It is that spirit of comaraderie and esprit de corps which has spurred the completion of this account and which has justified the telling of the 39th Fighter Squadron's story.

CONTENTS

ACKNOWLEDGMENTS

An effort like this is difficult enough under the best circumstances but there are some people who make a great difference. The World War II portion was helped enormously by veterans of the 39th, especially Col. Charles King who gave more generously than I could ever have hoped. Other 39th people who gave valuable assistance included Stanley Andrews, Charles Covey, Curran "Jack" Jones, John Lane, James Querns, Richard Smith and Ross Whistler.

Other people who came to the aid of the effort included Bob Brooks, Dick Hill, Bruce Hoy and John McCoy of the George AFB History office. And a special thanks goes to Jeffrey L. Ethell who came through once again.

Cover photo courtesy James Querns

Cobra in the Clouds

The Bell P-39 Airacobra, the type the 39th Fighter Squadron took into combat in 1942. It coined the squadron's "Cobra" namesake.

R.M. Hill

39th F.S. Formation

On December 6, 1941 the day before the attack on Pearl Harbor, the 39th Fighter Squadron had just completed maneuvers and settled in at its new base at Baer Field, at Fort Wayne, Indiana. The P-39 Airacobras with which the squadron was equipped were older models and were not fully suited for combat training. At one point all pilots were put to work removing the six hundred Phillips-head screws on the upper wing panel to facilitate installation of leak-proof tanks. Armament was also incomplete; a steel bar served for the 37mm nose cannon and air-to-air gunnery training was non-existent. Thus, on the eve of all out war, a grave disparity existed between the preparedness of the 39th Fighter Squadron and its prospective Japanese enemy.

There was one advantage the young pilots of the 39th did enjoy. Although they would find in the near future that the Japanese possessed a murderous edge in aircraft and tactics, the pilots of the 39th were blessed with eagerness and youthful optimism. This spirit would help them through their first combat experiences and set a pattern leading to the establishment of a proud record.

Charles King was fortunate enough to get a two day pass and a ride to his home in Columbus, Ohio. This flight of the 39th, which was commanded by Bob Faurot, was the first to land at their new base in Indiana and King took the opportunity for a weekend break. Since he had been trained for war and had been expecting it, young Lieutenant King was not really surprised at the news that came over the old Crosby cathedral radio that Sunday but was stunned at the bold Japanese venture at Pearl Harbor. He was emotionally detached when he received the telephone call summoning him back to his unit; it was an unnecessary exercise. When King left his home this time he would not return until he had faced two perilous years of combat, which included command of the 39th Fighter Squadron.

Frank Royal and Curran Jones were also off that weekend, flying from Baer Field to Selfridge Field, Royal to see his girlfriend and Jones to be best man at the wedding of a P-38 pilot stationed at the base. Royal called Jones at the wedding reception as soon as he heard the news of Pearl Harbor. The pair hustled back to their planes and returned to Baer Field. Jones had had some champagne at the wedding reception and felt like a tiger on the way back. He planned some personal retaliation for the Japanese military machine. His exhurberance for combat would never again be so high although he was destined for some excitement in the coming months.

Pilots of the 39th were champing at the bit while awaiting overseas orders. The 31st Fighter Group to which the 39th was originally attached deployed to England where they received Spitfire aircraft and participated in the North African landings . . . minus the Cobras of the 39th. The disconsolate Squadron pilots had to wait for another two months before they left the U.S.

Personnel of the Squadron boarded the USS *Ancon* in San Francisco bound for Australia at the end of January 1942. Arriving at Brisbane on February 25, the Squadron was quartered briefly at Ascot Race Track. Sailing again on the third of March, they reached Melbourne on the eighth where they were quartered until moving to Williamstown on April 3. The first operational base from which the 39th would fly was Woodstock in Northern Australia where the Squadron received its brand-new P-400 Airacobras on April 20, 1942, and attached to the 35th Fighter Group.

For the first time it was possible to bring the 39th up to combat strength. Conditions were rustic at best with tent living and operational runways hacked out like great Xs in the kunai grass and scrub trees of the coastal area. Some time was logged in aerial gunnery practice at last when pilots found that the shadows their Airacobras cast on the waters off the coast made perfect targets for deflection shooting practice.

Pre-war stateside maneuvers near Lexington, Ky. Dec. 5, 1941. L. to R. Lts. Greco, Lynch, Fink, Zimlich and Holloway. They were classified as the 39th Pursuit Squadron of the 31st Pursuit Group at this time.

R. Greco

Entering combat, May-July 1942

During April the Japanese Navy moved its best Zero pilots into Lae on northern New Guinea. Regular raids were made against Port Moresby in anticipation of overwhelming the Allied stronghold. The Japanese pilots encountered P-39 and P-40 aircraft which fared badly against the Zero. Part of the superiority enjoyed by the Japanese lay in the fact that the Zero was somewhat faster than the American fighters, had a better climb and ceiling and was infinitely more maneuverable. In addition, the range of the Zero was much greater than the American types, which gave emphasis to the offensive posture of the Japanese. It is known that exaggerated claims were made by both sides but a realistic tally indicates that the Japanese were beating the American fighter defense by a wide margin.

Into this bleak situation the neophyte 39th pilots were straining to enter. Training had dictated standard European tactics. That is, the American pilots would attempt to get above their opponents and make diving passes. Japanese strategy throughout the war depended on sucking American contingents into close commitment and defeating them in individual combat. Americans learned quickly the value of flexibility in adapting their tactics to those of the enemy as witnessed by the 39th's first battles.

The first contact with Japanese aircraft came on May 17, 1942, and an anonymous 39th pilot who kept a personal record of the early operations, wrote of the day:

"Lts. (Green), Adkins, Lynch, Wahl and Carey the first of our Sq. to participate in combat activities in New Guinea, arrived on the evening of May 16 and immediately reported for duty.

"Lt. (Green) leading a flight of 4 in 2-ship elements took off on an interception mission. When at an altitude of 11,000 feet a flight of 5 "0" fighters were sighted at the same level 90 degrees left flying in a Luftberry. Our flight maneuvered for position and attacked head-on. The enemy executed an Immelman giving them a position up the rear of our A/C and attempted pursuit. The result of the engagement was nil. E/A cowlings were painted red."

In point of fact, the first action experienced by 39th pilots was a volunteer affair that did not involve the unit as a whole. The five pilots mentioned in the diary eagerly took the opportunity to attach themselves to the 80th fighter Squadron for a temporary spate of action. One of the pilots, Lt. Tom Lynch, was destined to become a great ace of the Pacific war. He was also the rare sort of pilot who naturally took to leadership position. One of his traits was that he only claimed victories which were absolutely sure kills. His score would have been much higher if he had claimed every enemy plane that may have gone down before his guns.

Another remarkable fact was that he scored three of his victories in the Airacobra. Pilots who flew the P-39 were generally disenchanted with its performance, and the first operations of the 39th caused little enthusiasm for the airplane. But just what was accomplished and how the pilots reacted to the P-400s was an important factor in the record of the 39th. On May 20, Tommy Lynch scored the first victories for the Squadron.

Whatever the result of the combat, the 39th journalist recorded some telling attitudes about the weapon at hand:

"Lts. Lynch, Wahl, and Adkins, flying in a 5-ship flight in cooperation with the 35th Sq., 8th Pur. Group, intercepted 6 Zero fighters at around 15,000 feet. When first sighted, the enemy, who were flying loose echelon, observed our head-on attack. Lt. Lynch reported hits on 2 E/A but no results were determined. After attack E/A pulled around on our planes' tails and our pilots dived away. Lts. Wahl and Adkins reported no results. Enemy casualties were probably 2 Zeros. Our casualties had Lt. Lynch's plane shot up but landed safely. Lt. Carey bailed out of plane and was injured when he hit the ground."

Comments: Lt. Adkins: "Could have done better with a truck; it's more maneuverable and will go higher."

Curran Jones' controlled crash landing when main spindle on the landing gear broke just after touch down during stateside training. No one was hurt.

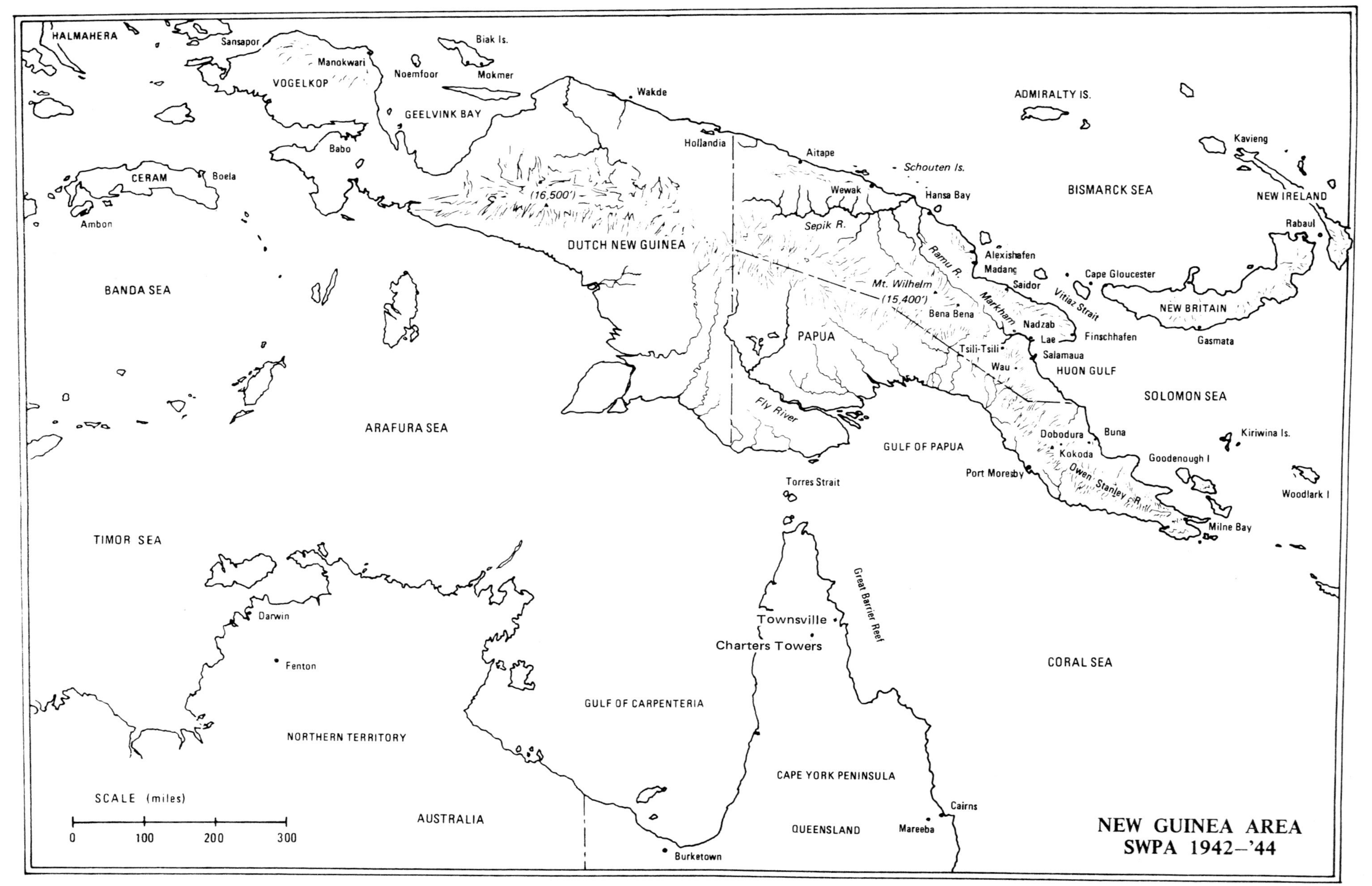
NEW GUINEA AREA
SWPA 1942–'44
NEW IRELAND
Rabaul
Kavieng
Woodlark I
Kiriwina Is.
Milne Bay
Goodenough I
NEW BRITAIN
Gasmata
SOLOMON SEA
BISMARCK SEA
Cape Gloucester
Vitiaz Strait
Finschhafen
HUON GULF
Buna
Dobodura
Kokoda
Owen Stanley R.
ADMIRALTY IS.
Alexishafen
Madang
Saidor
Nadzab
Lae
Salamaua
Markham
Wau
Tsili-Tsili
Port Moresby
CORAL SEA
Schouten Is.
Hansa Bay
Ramu R.
Mt. Wilhelm
(15,400')
Bena Bena
GULF OF PAPUA
Cairns
Mareeba
Great Barrier Reef
CAPE YORK PENINSULA
QUEENSLAND
Wewak
Aitape
Sepik R.
PAPUA
Torres Strait
Townsville
Charters Towers
Fly River
Hollandia
Wakde
DUTCH NEW GUINEA
(16,500')
GULF OF CARPENTERIA
Burketown
Biak Is.
Mokmer
GEELVINK BAY
Noemfoor
ARAFURA SEA
AUSTRALIA
Manokwari
Babo
VOGELKOP
NORTHERN TERRITORY
Fenton
Darwin
Sansapor
Boela
CERAM
BANDA SEA
TIMOR SEA
HALMAHERA
Ambon
SCALE (miles)
0
100
200
300

A P-400 (P-39), No. 19 "The Flying Arrow" in two-tone "British" camouflage. Southwest Pacific area.

King/Lane

Lt. Wahl: "Could have done damn good with altitude ship."

Apparently, Lynch was granted confirmation of two victories probably on the basis of wreckage identification of the Zeros in the jungle.

(At the time the Squadron scored its first victories, six other squadrons in the Southwest Pacific already had their first victories. Totals for those squadrons, as of May 20, were 7th Fighter Squadron 7 victories, 8th Fighter Squadron 17 victories, 9th Fighter Squadron 15 victories, 35th Fighter Squadron 12 victories, 36th Fighter Squadron 16 victories, and the 40th Fighter Squadron 1 victory. The first three squadrons made up the 49th Fighter Group, the next two were part of the 8th Fighter Group, while the 40th was part of the 35th Fighter Group, along with the 39th. The Group's third squadron, the 41st, was yet to score, as was the 80th Fighter Squadron of the 8th Fighter Group. Both would go operational in July 1942.)

For the next few days the New Guinea front was quiet for the 39th Fighter Squadron. Morale in the unit remained quite high, and rather than feeling anxiety at the prospect of combat in aircraft of limited capability, the pilots were simply frustrated at not being able to engage the enemy with greater ease.

Air activity increased on May 26. Lynch, Adkins and Wahl were flying again with the 35th Fighter Squadron excorting five transports to Wau. Sixteen Zeros popped up around the Mt. Lawson area southwest of Wau and the Airacobras broke to the right to take them on. Lynch fired, causing one Zero to disintegrate, while Gene Wahl sent another one down in flames and Adkins downed a third. The American formation suffered no casualties.

Once again the attitude of the 39th pilots was reinforced that better results were possible only with improved equipment. What they wanted most of all was an airplane that could get above the Zero and perform well at altitude. Airacobra operations would continue for the 39th through their first tour of operations, and the pilots would fight with enthusiasm if not with complete ease. June would see the Squadron operate as a unit for the first time.

May was nevertheless successful for the young 39th. Five victories were credited during the month, three of these falling to the P-400 of Tom Lynch. Another was credited to Frank Adkins who was a veteran of the early Java fighting and came along in time to offer his expertise. He had also scored one victory with the 17th Fighter Squadron in Java and would go on to further successes in Europe later on.

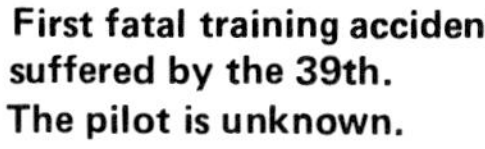

First fatal training accident suffered by the 39th. The pilot is unknown.

King/Lane

Only one loss was suffered by the 39th. On May 20 when Ralph Carey was forced to abandon his fighter during the battle. He was shot down while in a dive from 22,000 feet and broke both legs during his parachute jump escape. For some time after his life hung in the balance but, fortunately, he not only survived the ordeal but managed to walk again.

On the first day of June the 39th was ordered to relieve the 35th Fighter Squadron, and Major Jack Berry led eight other Airacobras to Port Moresby that morning. Other pilots in the flight included Lts. Bob Faurot, Charles King, Ralph Martin, Frank Angier, Walter Beane, Gene Wahl, Curran Jones (who probably felt a bit less ferocious than he had on that flight a few months before), Frank Adkins and George Parker. Adkins had some trouble with his airplane and was delayed but caught up with the flight at Horn Island. Unfortunately, when the flight was held up at Horn, Adkins contracted Dengue fever and was evacuated by B-17 to Townsville.

Weather kept the Squadron at Horn for several days. On June 3, Frank Royal came up from Woodstock with orders to proceed with Capt. Berry to Port Moresby. They arrived at 7-Mile Drome late in the afternoon. The rest of the Squadron flew patrols from Horn Island until Charlie Sullivan came up by Australian Hudson bomber on June 6, to inform the squadron that it would fly to Port Moresby in the afternoon if weather permitted.

There were still many difficult tasks ahead, however, and some harrowing incidents as witnessed by the June 7 entry in the journalist's diary:

"We are to work out of 12-Mile strip as soon as it is ready. A few planes are there now. However, until it is in condition some will be working out of 7-Mile in cooperation with the 40th Sq.

"Lt. Bartlett's plane (7150) failed on take-off from 12-Mile resulting in a crash landing. The plane actually turned end over end, cartwheel fashion, and was demolished. He was unscathed except for a small cut over his right eye.

"No enemy contact—some patrolling. News of the Midway Naval victory was welcomed joyously by the Sq. when the news came over the radio."

The entry for the next day described another accident which also ended happily, this time for Lt. Beane:

"At around 1500 Lt. Wahl took off and led a group of 8 off 12-Mile for a familiarization flight. Upon their return, Lt. Beane in landing hit hard and gave it the gun to go around thinking his undercarriage was damaged. His engine was missing (evidently something was jarred) and he was unable to gain altitude. Using flaps and full power he reached a height of about 400 feet, holding it in a stalling attitude all of the time. Suddenly the plane started to loose altitude and went into a spin. We noticed the chute stream out without opening at about 250 or 330 feet. Here's one for Ripley—Beane saw he was falling behind the plane. When it crashed it exploded, causing huge billows of smoke and fire to envelope him, and the chute to open. He gave a heave on the shroud lines and landed 25 feet from the inferno. Pilot received slight burns around the face and hair and a bruised right knee. The plane (349) was completely demolished."

An escort mission on June 9 gave the 39th a significant victory in combat. B-26s of the 22nd Bomb Group and B-17s of the 19th Bomb Group attacked Lae, and the B-26s came back with a hornet's nest of Zeros angrily buzzing for revenge. Among the Zero pilots were the foremost aces of the Japanese Navy—including Saburo Sakai and Satoshi Yoshino who was credited with at least fifteen American aircraft shot down.

King/Lane

Curran "Jack" Jones by his P-400 Airacobra about the time of the Cape Ward interception.

At least one of the B-26s was lost, but the veteran crews were holding their own when eight Airacobras of the 39th showed up at a point just north of Cape Ward Hunt. Joe Green, who was leading the 39th, heard the leader of the second flight ask for cover as he was going down to aid the bombers.

Curran "Jack" Jones had the perfect moment for his baptism of fire, and recently recounted his battle with the crack Zero pilots:

"When I got down and could see Zeros scattered like flies and started to pick one out of the covey, Joe called on the RT to say that his prop was out and he was going home. I knew my flight was with me so I pressed home my attacks. We all four fired at one aircraft which did the usual steep climb followed by a kind of hammerhead stall. As we came around in trail formation my flight was strung out. I heard Price, my number 4 man call the number 3 man, Bartlett, and ask him, 'Is that you, Bartlett?' When the reply was negative, I heard Price say, 'Uh-oh.' As I looked back, there were five of us in the flight.

"I made an extremely tight right hand turn and . . . passing my number 2, 3 and 4 headed back to engage the Zero which was stalking my number 4 man, Price. I still had a good bit of speed and was running wide open when the Zero started his usual vertical climb so that I was able to go up a good distance with him. I started firing short bursts. I realized my speed was getting low, but was too busy to be concerned. Most fortunately I saw what I think to be one of my 20mm shells explode in front of the cockpit.

Airacobra No. 94 carried typical personalized insignia on the cockpit door. . .here a roadrunner bird character carrying a machine gun. Some 39th aircraft did not conform to standard numbering range of 10 to 39.

C. King

Charles King in Airacobra No. 27 on patrol over Port Moresby area.

"By then the Zero had sort of flattened out and there was a movement of the pilot as I approached, and I realized that he was climbing out of the cockpit. I tried to pull in a little tighter in order to shoot him off the wing but would have stalled out as I was down, I believe, around 140 mph with wide open throttle. As I passed behind his crate, he was holding on to the cockpit and looking back at me and the nose of the Zero was just beginning to drop. He had no parachute. There were two red diagonal stripes just after the cockpit around the fuselage. I told the fellows that he looked at me as though I was the last man he would see alive.

"After clearing my tail I looked back and followed the Zero into the water. It smoked only slightly. I went on the RT in an excited fashion and said, "Joe, did you see that one go in?" But I got no answer. I looked for friendlies or enemies and I was totally alone in that big sky."

Five Zeros were claimed on the June 9 mission although postwar evidence indicates only one Japanese was lost. Other 39th claims for this particular combat are somewhat vague and seem to reflect the excitement of green pilots in their first real battle. There is little doubt that Jones claimed the only Zero to be lost in the battle–that of ace Satoshi Yoshino.

One other fact of interest emerged from this action. Years after the war Jones learned not only the identity of his victory but that one of the people in the B-26s was Lyndon Baines Johnson, then a Naval Lieutenant Commander on an inspection tour.

For the next few days the 39th was more harried by accident and Dengue fever than by the enemy. Gene Wahl made a forced landing on the ninth and wandered through head-tall grass before he was picked up by natives and returned to the 39th camp on June 15. On that same day Tom Lynch and Frank Adkins returned to duty from a bout with Dengue and were just in time for a rough mission on the sixteenth.

Early that morning, sixteen 39th Airacobras were scrambled from Laloki and 12-Mile drome, fourteen fighters eventually making a sweep but finding no enemy. Later in the day another 39th flight found better hunting which was recorded by the squadron journalist in the entry for the day:

"From Laloki, Lt. Royal led a flight of 4 to the Rigo areas climbing to 22,000 feet. Here he was joined by Lts. Rauch and Suehr. A formation of 18 Zeros were sighted directly above at 25,000 feet proceeding in a southwesterly direction. A few Zeros evidently broke away from the main formation and maneuvered for attack. 1 E/A got on the tail of Lt. Rehrer without apparently being observed and Rehrer's plane was seen to go down in flames–none of our pilots saw a parachute open. Lt. Faurot turned into this attacking Zero, attempting a head-on attack–no results were noted and he received bullet holes in his A/C. Lt. Lynch was flying alone below and behind, attempting to catch Lt. Royal's formation, when he sighted and attacked 2 E/A. Four other E/A immediately joined the melee, shooting Lynch's plane up badly. He dove and lost the enemy and headed for 7-Mile. At about 7,000 feet the plane's engine blew up, and with this he headed for a forced

Charles Sullivan by his P-39; carried diving-eagle art work on the door.

landing on the beach. However, seeing this was impossible, the pilot bailed out at around 800 feet over the bay. The plane exploded when it hit the water. Lt. Lynch landed safely and was picked up by a native in a small boat. His right arm was broken—and we lost a damn good pilot."

On the mission Lynch happened to be wearing a Mae West life jacket, which was new equipment to the 39th. When other pilots asked him how his vest worked, Lynch replied that he didn't inflate the thing because he remembered the shark stories about the area and didn't want anything to interfere with his swimming! Lynch took the experience in his stride and returned to action in a few weeks. One Zero was claimed by Royal, and Rehrer reportedly returned to the Squadron after a trek in the jungle. The next day, June 17, eleven 39th Airacobras were scrambled to intercept a formation of G3M Nell bombers covered by a number of Zeros. Three P-400s were able to make contact with the Japanese formation but Lt. Charles King, the leader of the American flight, had difficulty sighting the enemy. As he called to Lt. Suehr and told him to take the lead in attacking the enemy formation, King inadvertently flew through the front of the bomber group. The Zero escort frantically angled to intercept King's fighter but only managed to zoom past his tail in wild dives.

King delivered a talk to air cadets later in the war and told the story on himself:

> "As soon as they passed me I used coordination 'Stick and Throttle forward'. In my pushover, dirt from the cockpit floor was flying all over the place and the accumulation of sweat and condensation in my old style oxygen mask ran down my face. I was convinced the bombers plus the fighters had filled the cockpit with lead and my blood was being given for the Fatherland. As far as I know, they did not attempt to follow me in my dive. At about 12,000', I pulled out and assumed a course parallel with the bombers. I climbed and finally one black dot appeared and as I edged closer, others came. I figured I had had enough odds against me for the day and went home.
>
> "Then I learned I had had only one man with me. The 3rd man had lost us in a turn. Lt. Suehrs had attacked the bombers with damaging results. He got away unscathed. He asked me if I was brave or just crazy. When I told him I was just blind, he just about passed out. Piecing it together I figured I had missed them by use of poor searching procedure. I looked above and below the horizon and they were right on it and were against a blue-green sea background. Live and learn—if you're lucky and a bit quick."

The remainder of June was a repetition of earlier missions. Time after time, formations of approximately eighteen Japanese bombers escorted by Zero fighters would raid Port Moresby. Five 39th pilots were shot down during the month for claims of nine Zeros. It should be noted that good fortune attended the Squadron during the early operations and in spite of its aircraft losses all downed pilots returned alive. Five other aircraft were also written off in accidents.

However dangerous aerial combat may have been, there were also humorous events. An entry in the Squadron diary for June 21 illustrates this:

> "For the past 2 days Laloki has been subjected to strafing attacks not by the amber men but by our own Lt. Royal, forgetting to turn the gun switches off. Bullets clipped the runway both times when he made his landing. Extra slit trenches are now in the process of being provided by our personnel."

Scrambles and patrols continued throughout the first few days of July without much success in contacting raiding Japanese aircraft. A number of bombing missions and even a commando raid were made against Lae and Salamaua which probably dampened the Japanese effort. On the night of July 3, the Japanese did make one unsuccessful attempt to bomb 7-Mile Drome, however.

The Fourth of July was a rough one for the 39th. Around 10:00 in the morning an alert was sounded and thirteen planes took to the air, led by Major Berry. About forty miles north of 30-Mile Drome, the Squadron was bounced by Zeros.

Charlie King was in the flight and remembers breaking to the left as one Zero came down in a hard dive on his tail. He half-rolled and fired straight down at the enemy plane when it roared by. He lost the Zero once it zoomed away, and he went for the cloud cover over the mountains. The electrical system and propeller went out in the Airacobra and he made a hot landing back at the base.

Others in the flight were not so fortunate. Lt. Price, in Major Berry's flight, was attacked by four Zeros and his plane was badly damaged before Lt. Martin was able to rescue him by forcing the enemy to break away. Major Berry and Lt. Sullivan were able to elude their pursuers by diving into the clouds. Lt. Royal received some damage to his aircraft when he tried to dive away some Zeros circling another 39th pilot in a parachute.

When the scattered American formation finally returned home three of their fighters were missing; those of Angier, Marlett and Foster. Other Airacobras were badly damaged, and

Damage caused by enemy bombers on 7-mile Drome.

C. King

Gene Wahl with his "Wahl Eye II." "Eight Ball" insignia decor on cockpit door. Gene's Airacobra No. 13 was one of the first to employ the shark mouth motif in the 39th F.S.

C. King

not one Japanese fighter had been claimed. It was the worst defeat the Squadron would ever suffer.

July took much the same character as the preceding month in that patrols were maintained against the possibility of Japanese raids. By the 12th of July the three pilots shot down on the 4th were accounted for as safe and sound. Lt. Angier was the man in the parachute whom Lt. Royal tried to save from strafing. After the Zeros had attacked him as he drifted down, and also on the spot where he landed, he managed to contact some natives from a nearby village who guided him home. Lt. Foster was the only one of the downed pilots who required serious medical attention. He was injured during his parachute jump.

By the morning of July 22, the Japanese had landed on the east coast of New Guinea and were directly threatening Port Moresby. B-25s, probably of the 3rd Bomb Group, plus Airacobras of the 39th and 80th Fighter Squadrons bombed and strafed the shipping off Gona, just north of Buna.

Bob Faurot and Charles King led a section of the strafers down on the landing barges with little effect as the landings were completed in the afternoon. One Airacobra pilot was hit and bailed out. Later in the day other missions were carried out against the landing beaches. One other A-24 escort was made to Gona by the 39th on July 24.

The last P-39 mission flown by the 39th also came on July 24. Like many other missions flown during the preceding two months, eight 39th fighters took off and patrolled without seeing any Japanese. From July 25 through the end of the month, 39th personnel moved back to Townsville as the 80th Fighter Squadron moved in to relieve them.

If anything epitomized the 39th's first tour of combat it was that circumstance battled the Squadron as effectively as did the enemy. Although several members of the squadron did succumb to the severe phychological pressure of the battles they faced, the spirit of the organization was high, and the pilots remained eager to return to combat in more suitable aircraft. They would not be disappointed.

P-400 Serials of the 39th

AC 332	BX 147
AC 348	AC 374
AC 373	AC 361 Lost 18 June 42 (Greene)
BX 172	AC 345
BX 171	AC 347
BX 173	AC 103
BX 169 Lost 18 June 42	BW 163
BX 144	BW 180
BX 170	BW 153
BX 163	BW 137
BX 168	BW 169
BX 166	BW 176
BX 148	

9 P-400 shot down between 20 May and 4 Jul 42
7 P-400 damaged in action between 20 May and 4 Jul 42
8 P-400 lost on ops between 6 Jun and August 42

Capt. Robert Faurot's P-400 at 7-mile strip, June/July 1942, nicknamed "Poison."

R.M. Hill

Pilots took to the new P-38 like fish to water. Buzzing the field in exhileration.

Australian War Museum

Transition to the P-38, Aug.-Oct. '42

It was actually no more than a flip of the coin that determined the 39th Squadron would be the first P-38 unit in the Pacific Theatre. T/O 1-37, dated 1 July 1942 authorized the reorganization of an existing fighter squadron into a twin-engine unit. The 39th ground crews who had remained in New Guinea to service 80th Squadron aircraft, returned to the mainland on August 12.

Major Jack Berry was killed in a training accident during this period–the only 39th fatality during operations with the Airacobra. He was practice-bombing on an old hulk in an Australian harbor when his bomb exploded directly beneath his fighter. He was in turn replaced as squadron commander by Lts. Stephen David and Frank Royal prior to P-38 transition.

New pilots for the 39th were being assigned. During August about a dozen of them arrived, including Dick Smith and Stanley Andrews with considerable P-38 time accumulated. Another pilot ferried a P-38 to the 39th sometime in November and stayed with the Squadron long enough to begin his record as the top American air ace of all time–Lt. Richard Ira Bong.

Meanwhile, the veterans of the 39th were luxuriating in the questionable comforts of old rustic Townsville. Most of the pilots sported full-growth beards, Charlie King was even mistaken for "Whiskers Blake", a notable Australian wrestler. During their interlude in Townsville, personnel learned that the Japanese move on Port Moresby had been halted, due in great part to tireless efforts by the Allies at Milne Bay.

Initial transition to the P-38 was impeded by some mechanical difficulties, including leaks in the cooling system and inexperienced maintenance. Stan Andrews recalls one P-38 that he picked up at a depot had its ailerons reversed; the right aileron was on the left wing and vice versa! Pilots were eager to get back into combat and were understandably anxious about transitioning to twin-engine aircraft. Improved components were rushed to the theatre to hasten operational readiness.

Actually, training in the P-38 was haphazard for the 39th. Curran Jones recalls that it was probably Bob Faurot who checked him out by telling him where all the switches were. Jones was admonished to slam the big airplane back on the runway if an engine cut-out on takeoff.

On one patrol during the training period, Jones found himself alone high over the Owen Stanleys when his wingman turned back with a malfunction. Jone's pitot head froze. He cursed helplessly when he couldn't find the heater switch and had no idea of his airspeed. Fortunately, the tube became unfrozen at a warmer altitude and a safe landing was made. Jones spent an hour in the cockpit after landing and finally found the heater switch behind the control column.

C. King

Charles King sports the beard he grew during the period after the squadron's first tour of duty.

USAF

Capt. Robert Faurot, January 20, 1943, by his famous P-38.

Charles King also had a spartan introduction to the P-38. On September 23 he made a forty-five minute check flight over Ross River in a P-38E. The next day he repeated the flight, and on October 3 he flew a two hour sortie to complete his transition. New pilots in the squadron with P-38 time did what they could to help, but the fact was that the 39th would enter combat with minimal experience on the P-38.

By the 10th of October the 39th was moving once again to an operational base. Staging through Horn Island was repeated, and the Squadron settled in at 14-Mile Drome. The Japanese were quick to spot the P-38, for there are records of P-38 sightings by Japanese naval air units in October. There were no contacts, however, since the 39th was not able to find any significant opposition throughout its autumn 1942 patrols.

It may be useful to look at entries from the flight record of Lt. Stanley Andrews for an idea of the 39th's preparation for battle:

10/18/42	Over Wingaby
10/23/42	Up and down S. E. Coast
10/27/42	Up and down S. E. Coast
10/29/42	To Milne Bay
11/1/42	Milne Bay to Laloki
11/4/42	Patrol to Salamaua
11/8/42	Laloki
11/9/42	Laloki

One distinctive marking was adopted by the 39th in October. Red sharkmouths with gleeming white teeth were painted on the engine cowlings beneath the blue propeller spinners. Charles King for one was not pleased with the flamboyant motif, but the general feeling was that now the Squadron was certainly identifiable in combat.

Thus, the 39th Fighter Squadron (Twin Engine), as it became known on November 6, was ready for battle once again. Only this time it was ready with a vengeance.

Air battles over Lae, Dec.-Feb. '43

There were two notable events in 39th Fighter Squadron history during November 1942. The first was the flight of eight P-38s to Guadalcanal on November 13. Bob Faurot led the flight which consisted of King, Baker, Denton, Haigler, Lane, Shifflet and Sparks. Although they were shelled upon landing by the Japanese, they were treated well by the other squadrons operating from Guadalcanal. The eight pilots stayed until November 22 when they returned to New Guinea minus a P-38 that developed mechanical trouble and was subsequently transferred to the Thirteenth Air Force.

The other event was more unusual and involved the first P-38 victory in New Guinea . . . albeit not a shot was fired. There is sufficient documentation that on November 26 Bob Faurot led four P-38s to dive-bomb Lae airfield and knocked a Zero from the sky with the explosion of his 500 pound bomb!

The P-38 Lightning in trim, undecorated form as used by the 39th Fighter Squadron.

Frank Royal, John Lane and Curran Jones also saw the incident and Jones wrote of his eyewitness account:

"We echeloned to the left and approached Lae from the northwest (?). We dove in a fairly shallow dive and released on Faurot's estimation, as flight leader, two 500-lb bombs each.

"As we made a gentle climbing turn to the right I could see Zeros taking off and did witness Bob Faurot's bombs go off at the end of the runway and a Zero make a 180-degree turn and pancake in the water. I think we had some chatter on the radio about that and of course we kidded Bob after we got back. We were not intercepted."

Faurot was given a medal and full credit for the victory.

December was a month of patrols and practice missions with little contact with the enemy. A puzzling fact is the existence of Japanese records indicating that American P-38s were engaged when no records are evident for the latter part of the year from American sources. That is, no engagements until two days after Christmas 1942, when the 39th Fighter Squadron began war in earnest.

December 27 started for the 39th with a scramble in the morning around the Buna-Gona area. Twelve P-38s were in three flights of four at 21,000 feet over Dobodura when P-40s at a lower altitude called that they had encountered a formation of Zeros, D3Y Vals and a new fighter to the New Guinea front, the Ki-43 Oscar.

Tom Lynch sensed a big battle at hand and led his wingman down on the Oscar formation. Lynch's enthusiasm was obviously high for he fired at three of the enemy, shooting one almost in half. He then tangled with another group of Oscars and one was observed to crash into Huon Gulf. When his ammunition was exhausted Tommy flew back at full speed to 14-Mile Drome, jumped out of his P-38 and urged the astounded ground crew to prepare another Lightning because "the sky is full of Japs over Buna".

Meanwhile, Dick Bong and Ken Sparks dived on the formation of Vals which was covered by a Zero escort. Bong shot at a Val which exploded then fired a short burst at a Zero which Bong's wingman saw fall into the sea. Bong then tried to attack three other dive-bombers without avail.

Sparks and his wingman attacked a group of Zeros and one fighter went down smoking under Sparks' guns. As he attacked another aircraft, however, he looked back to see five Zeros blazing away on his tail. He was able to dive away from the pursuers but not before an engine was shot out. When he attempted to make a landing at the new strip at Dobudura two Vals got in the way. He fired at one and sent it disintegrating into the sea.

The third flight of P-38s waded into a gaggle of Zeros and Stanley Andrews and Hoyt Easton jumped a group of three that broke in different directions. Andrews took the one on the left and fired at close range until the Zero's fin and rudder fluttered off and it spun into the water. Eason took after one that turned to the right and fired a burst into the cockpit, sending the Zero into the water, also. Eason shot another Zero off the tail of a P-40 which then turned and fired at the Japanese until it hit the water. The P-40 pilot confirmed the Zero for Eason, partly out of gratitude, no doubt.

Other 39th pilots were credited with victories during the combat. Lt. Carl Planck shot one of five Zeros off Lt. Charles Gallup's tail. Gallup fired at a Zero which had overshot his P-38, and the Japanese was observed to crash. Gallup turned like mad to break combat and finally lost his attackers. The P-38s finally made for home as the combat subsided.

Final credit for the mission totaled seven Zeros, two Oscars and two Vals confirmed and two Zeros and one Oscar probably destroyed. Japanese records do not support the extensive claims made by the Americans, and the fact that most of the pilots were in their first combat suggests that exhuberance was higher than objectivity. There is not any doubt, however, that the 39th emerged victorious; Sparks flew the only P-38 to suffer serious damage that day.

Another chance for the 39th to shine came on the last day of 1942 when A-20s, B-25s and B-26s were escorted to Lae. The P-38s were at 15,000 feet near the target when a dozen Zeros were sighted far below. The P-38s dived like stones into the attack. In the first pass Lt. John "Shady" Lane saw a Zero exploded by Tom Lynch and then shot down another himself in the head-on attack.

Hoyt Eason led his flight down into the formation of eager and skilled Japanese. He shot a Zero off the tail of another P-38 and the enemy fighter was observed to crash on the end of Lae airstrip. Another Zero fell to his guns and crashed about three miles from Lae. Eason downed yet another Zero and was the first 39th pilot to claim three enemy aircraft in a single combat.

Ken Sparks caught one Zero in a slow roll, and it was observed to crash in the woods a few miles northeast of Lae.

Sparks met another determined enemy in a head-on pass and the two planes brushed each other, Sparks suffering a damaged wingtip and the Japanese going down minus part of his wing. Dick Bong experienced a frustrating day, firing at no less than six Japanese and claiming only one probable.

Once again the claims may have been overly enthusiastic, but nine Zeros were officially credited to the Squadron and no more than three P-38s were damaged, Ken Sparks again the worst casualty with a missing wingtip. Tom Lynch had increased his score to seven confirmed, and Hoyt Eason was the first recognized P-38 air ace of the Southwest Pacific with five confirmed victories. Morale was fantastically high and the 39th was beginning to experience its greatest period of fame.

Even with their relative inexperience with the big fighter the 39th did exceptionally well. For the first time in the Pacific fighting, American pilots had an airplane that could outclimb the Zero under most conditions. The Lightning was faster than the Zero, usually had the advantage of altitude and possessed the devastating firepower of four .50 calibre machine guns and a 20mm cannon concentrated in the nose. In the months to come, the combination of skill, eagerness and chine guns and a 20mm cannon concentrated in the nose. In the months to come, the combination of skill, eagerness and an effective weapon would make the 39th the premier fighter squadron of the Fifth Air Force.

On January 6, 1943, the 39th returned to Lae for a disappointing dive-bombing mission on a Japanese convoy in the area. After the bombs were fruitlessly dispensed, a formation of Oscars engaged the P-38s. Old hand Curran Jones added two Oscars to his bag, and seven other pilots were acredited with a Japanese fighter each. Once again no P-38s were lost. The next day another strike was made at Lae. Again the Oscars intercepted and lost seven of their number to the 39th, two to Bong. No P-38s were lost.

Tom Lynch got one of the Oscars on this mission for his eighth confirmed victory. He also had the experience of claiming an enemy ship. A heavy bomber observed Lynch's run on the Japanese transport and the ship burst into flame as Lynch withdrew. The vessel blew up as the bomber went in to finish it.

January 8 produced a climax to the 39th's first round of action with the P-38 when the Squadron escorted bombers to Lae again and made claims of sixteen Oscars destroyed in the air, one by Bong. This total, in fact, represented a record number of aerial victories for the Fifth Air Force fighter squadrons up to that point of the war. The 39th would never exceed that total of claims for a single combat day.

There were actually three escorts involved. The first took Beaufighters and B-25s to Lae and four Oscars fell to the Squadron. The second escort took B-25s and B-26s back to Lae and five more Japanese defenders fell to the 39th.

Seven Oscars were credited to the 39th on the third escort which took B-17s and B-24s to the beleaguered Japanese base. Dick Suehr claimed two Oscars but was bothered by one persistent enemy fighter who insisted on attacking the heavily-armed American fighter head-on. Dick Bong had sprayed some ammunition without effect among the Oscars when he noticed Suehr under attack and obliged him by shooting the pesky enemy down. It was Bong's fifth victory.

Ken Sparks managed to claim an Oscar on the first and third missions. During the first escort, he managed to shoot an Oscar off the tail of Lt. Kish. At the time, most American pilots believed the Oscar was a version of the Mitsubishi Zero. His narrative is revealing of the figher pilot's state of mind during this period of the war:

> "On our first pass I knocked down one Zero off of Lt. Kish's tail. He went down and out, I saw about a yard of orange red flame shoot out from behind his cockpit. I saw my flight leader, Captain Faurot, shoot down one enemy plane on that first pass. In addition to this I saw three Zero's, my own, Capt. Faurot's, and another hit the water when I pulled up to gain altitude. Later I saw one hit the water west of the convoy. I fired on my probable toward the end of the fight."

The 39th suffered its first casualty in P-38s when Lt. John Mangas failed to return after the third mission.

A period of inactivity followed the Lae operations and the 39th could tally the score. As of Janauary 8, 1943, 65 Japanese aircraft had been claimed by Squadron pilots since the 39th began keeping score on May 20, 1942. Totals for other

Australian War Museum

The overworked, long neglected and little publicized aircraft mechanic work on P-38s of the 39th F.S. Above; guns get a checking over. Left; crew working on port engine. Polished oval spot just above shark teeth is bare metal area which formed a mirror, giving the pilot a visual check on nose gear.

squadrons were 46 Japanese aircraft for both the 8th and 9th Fighter Squadrons, while the 7th and 36th Fighter Squadrons' tallies stood at 32 and 26.

From May 1942 through January 8, 1943, no more than ten 39th aircraft had been lost in action, including Mangas' P-38. Some damage had been done to surface targets, such as Lynch's sinking of the transport, but in general the P-38 did no fare very well as a dive-bomber in the minds of 39th pilots.

In any event, Tom Lynch had become one of the top aces of the Southwest Pacific with eight victories. Sparks followed with seven while Eason had six and Bong and Gallup had five.

By the 1st of February 1943 the 39th was the highest scoring fighter squadron in the Fifth Air Force. The 9th Fighter Squadron followed closely with forty-six victories (seven by three of its pilots flying with the 39th) and would offer even stiffer competition when it entered combat with the P-38 in February. Other units (namely the 80th Squadron) would enter combat and perform brilliantly with the P-38 in the months to come thanks in large part to the fighting example provided by the 39th Fighter Squadron.

There was only one major action in which the 39th was involved during February and that was an interception near Wau on the 6th. Fifth Air Force fighters claimed twenty-three Japanese that day of which the 39th accounted for one Oscar destroyed and one probably destroyed. Claims by 39th pilots were becoming more conservative with experience in battle.

Battle of the Bismarck Sea, Mar. '43

Dick Bong was moved back to the 9th Fighter Squadron permanently in January to help with their conversion to the P-38. He helped them get into shape in time for their role in the air battles over a large Japanese convoy trying to reinforce Lae early in March. The two P-38 squadrons would claim more than twenty enemy fighters during the destruction of the convoy of at least eight freighters and eight destroyers, in a battle which became known as the Battle of the Bismarck Sea.

Six thousand troops were being hastened from Rabaul to Lae during foul weather which the Japanese hoped would cover the convoy from Allied air attack. Fewer than a thousand troops would ever reach Lae. Most would go to the bottom of the sea during the battle.

It was on March 1 that the convoy was sighted by a Liberator bomber, and over the next three days B-17s, B-25s, A-20s and Bristol Beaufighters sank all the transports and strafed survivors found in the water. The grim order had been given to annihilate the enemy force and the aircrews reluctantly obeyed.

During the second day of March, the 39th was over the convoy in rough weather. The P-38s ran into a trio of Oscars, and Wil Marlett in Jack Jones' flight beat everyone to the punch by sending one of the Japanese down in flames. Charlie King then aimed his fighter at an Oscar which was climbing on the verge of a stall and sent it spinning down into the clouds trailing heavy black smoke.

The next day thirty-two P-38s of the 9th and 39th Squadrons escorted bombers in the morning, and eleven 39th aircraft flew a second mission in the afternoon. More than fifteen Japanese were claimed by P-38s during the day, eleven by the 39th. Paul Stanch began his scoring by downing two fighters as did "Jack" Jones to become the newest Squadron ace, and Tom Lynch claimed a Zero for his ninth victory. Maj. Prentice, the Squadron commander, Stanley Andrews, John Lane,

R.M. Hill

Major George Prentice, C.O. of 39th F.S. and later C.O. of the 475th Fighter Group.

Henry Turick, Richard Smith and Charlie Sullivan all scored single victories.

In point of fact, Stanley Andrews claimed his victory, an Oscar, on the only flight he made during the battle. He recently recorded some of his memories of the day:

"The situation as far as I was concerned really started the night before because there seemed to be a lot of confusion as to whether we would fly cover or go in for dive bombing. I was pretty lousy at dive bombing, about all I ever did was get near misses that helped wash the ships down. Anyway, the night before, I recall that we (the pilots) were out helping the crew chiefs put on bombs and then take them off to put on tanks. This seemed to have been changed a couple of times before the night was over.

"On the 3rd of March it was quite a gaggle . . . it was a beautifully clear day as you would frequently see in that area, you could see the coast of New Britain to the north and off to the west you could see the area of New Guinea around the tip of land near Finschhaven. The convoy of ships were spread out ahead and below us making white wakes in the deep blue sea. Had it not been such a terrifying situation you would call it ethereal. We, (39th) along with other fighter outfits, were flying top cover for just about anything that could drop a bomb. We were about 20 thousand (feet).

"As I recall, we were flying high and to the rear. The bombers that were down on the deck skip bombing were first since they could sneak in while the high planes to the rear were to cause any low flying Japs to start climbing to intercept us. Well, they didn't have to climb much, it looked like a swarm of bees coming at us from probably 15

Through the window of a Dodge truck, P-38, No. 27 with command stripes on aft boom, runs up the engines for takeoff. Squadron Commander Charles King at the controls. C. King

James Walter's P-38, No. 19 parked near Port Moresby during lull in air action. J. Lane

L. to R. Lts. John Dunbar, unknown, Edward Randall and James Walters, early 1943. J. Lane

Informal gathering of 39th pilots after a mission. It is surprising how many have cameras. Officer to right, looking at photographer, is believed to be from Group Headquarters. J. Lane

Posed in front of Tom Lynch's P-38, No. 10, L. to R. kneeling; Charles Sullivan, Thomas Lynch, Kenneth Sparks, standing; Richard Suehr, John Lane, Stanley Andrews. J. Lane

P-38, No. 15 "Thumper", assigned to John Lane, in which Dick Bong scored his first victory on Dec. 27, 1942. L. to R. F/O W. Beane, an Australian photographer, Tom Lynch, Dick Suehr and John Lane. King/Lane

thousand feet. About this time, the first skip bombers were hitting and the explosions on the surface started up. Fires, splashes, black smoke and then we were into it with the Japs. I lost most interest in the sea situation and concentrated on the air situation. The ones that I didn't see riddled

S. Andrews

Stanley Andrews by his P-38, No. 22, nicknamed "Lil Woman 2nd."

my plane but didn't hit the coolant systems or engines. In fact, I didn't know that I had been sieved until I landed . . . 93 holes. (I didn't count them, my crew chief did). Both tires on the main gear were shot out. Needless to say, I didn't roll very far. The plane was salvaged as I recall."

Casualties for the 39th were heavy that day, three planes and pilots being lost. Other pilots in the Squadron claimed that Bob Faurot was just worn out and could not put on the fine edge required in air combat. He noticed a B-17 at low altitude beset by a number of Japanese fighters. His flight, which consisted of Lt. Eason and Lt. Shifflet, went down to the aid of the bomber, and all three Lightnings were lost along with the B-17. The fate of the P-38s is not entirely clear, but one was seen to ditch, another disappeared into a cloud and a parachute was observed drifting toward shore. Unfortunately, none of the aircrew ever returned.

As the mopping up continued on March 4 and 5 the 39th participated with escorts and dive-bombing missions. The dive-bomber role was quickly fading for the Squadron since the P-38 was too fast in a dive for efficient bombing. Ken Sparks accounted for two more enemy fighters on the 4th and Paul Stanch and Ben Widmann each claimed another.

Along the New Guinea coast, Apr.-Sept. '43

Major George Prentice relinquished command of the 39th on March 24 to take over the soon-to-be-formed 475th Fighter Group—the first all-P-38 group in the Pacific. Replacing him was Major Tom Lynch who was one of the leading Fifth Air Force aces at the time with nine confirmed victories.

Curran Jones was a good friend of Lynch and remembers a time when Lynch reprimanded him during the Bismarck Sea battle. Jones had warned other pilots over the radio not to fire anymore at a Japanese fighter he had just sent down smoking. Later, Lynch told him in no uncertain terms that nobody is going to stop shooting at an enemy just so somebody else can get the credit.

On another flight, Lynch chewed-out a pilot who had just flown under a railroad bridge, and the pilot seemed to fly away in humility after the radio tongue-lashing. Jones was surprised to see Lynch fly under the same bridge and decided to do the same thing. Later, Lynch smilingly rebuked Jones for following his example.

April saw a resurgence of the Japanese bombing effort against Port Moresby. The twelfth saw a frantic effort by Fifth Fighter Command when a large formation of bombers came down to bomb Port Moresby. The 39th managed to engage the formation which was covered by Zeros. Smith, Gallup and Charlie Sullivan all claimed G4M Betty bombers, and Harvey Clymer got one of the Zeros. But the big show of the day was made by Dick Suehr when he exploded a Betty over thousands of cheering witnesses at Port Moresby for his fifth victory.

The sping of 1943 was generally quiet for the 39th. Only about fifty claims were made by all Fifth Fighter Command units during May and June. Only two of those victories were scored by the 39th, and both of those were claimed by Tom Lynch.

It was the privilege of the 39th to operate directly under Fifth Fighter Command control since it commenced operations with the P-38 fighter. It would operate in virtual detachment from the parent 35th Fighter Group throughout 1943 until all three squadrons of the group converted to the P-47 and were reunited at Nadzab, New Guinea in December. Tom Lynch would provide leadership for the 39th until he returned to the United States in September on temporary duty.

The effectiveness of Lynch's leadership would be evident on May 8 when he scored the Squadron's only victory for the month. Sixteen P-38s were sent out to cover B-25s that were trying to attack a convoy three miles off the coast of Madang. Five Japanese fighters appeared and the P-38s drove them off, damaging one Hamp. Just as the fight was about to start, Lynch realized he could not drop his external tanks and radioed Charlie King that he should take the lead. When he turned to go home, however, he noticed that the Squadron was following him so he resumed command and went into the battle with a hung tank. Lynch sighted one Hamp late in the fight about 10 miles from Saidor at 1000 feet and sent it down for his tenth victory.

Maj. Tommy Lynch admires his credentials, 16 enemy aircraft downed to become highest scoring ace in the 39th. R.M. Hill

July began the next period of sustained air victories for the 39th. Scoring began modestly enough with two confirmed and three probables but perhaps more indicative of the intensive activity on July 18 was the experience of it's pilots.

Lts. Walter Baker and Lloyd Shipley were in John Lane's flight of six P-38s covering a flight of transports returning from Marilinan. During this combat the 39th was introduced to the Ki-61 Tony which bore a resemblance to Italian and German inline-engine fighters. Shipley mistook a Tony for a German fighter which is noted in his report:

> "We flew into Salamaua area at 20,000 feet just under the overcast and sighted part of the enemy midway between Lae and Salamaua at a low altitude, headed toward Lae. We dived toward them, dropped our tanks, but chandelled to attack four enemy who had dropped from the overcast. I shot at what I believe was an Me 109. It was a fleeting 90 degree deflection shot which I believed missed completely . . . I dived to attack a Zeke who saw me coming and turned into a head-on attack. I do not know if I was hitting him or not. While in the dive I noted several single engine fighters among several twin engine aircraft headed south. I then followed a single P-38 which was attacking three Zekes. Two of the Zekes came up into hammerhead stalls and one of them offered me a perfect sitting shot, but none of my guns were firing. I dived out to the south, charging my guns. I got 3 of my 50's firing and returned to the area. I tailed a P-38 into a head-on attack on 2 Zekes. I was a good way behind and the enemy did not see me. One Zeke refused the head-on attack and dived down and out to a 180 degree. I dived from 1000 feet above him and inside of his turn. I closed to about 600 feet and began firing. My tracers went ahead of him and it was then that he sighted me. He took no evasive action except to increase his dive and rate of turn. I steepened my dive and pulled my tracers into him. He kept diving and rolled onto his back as I went by him. The dive had taken me close to the top of hills and I do not believe the enemy would have had enough altitude to complete a split "S".

Shipley then saw a Japanese fighter making a desperate head-on attack on another P-38. Rather than a dead-straight course, the Japanese craft was nosing up and down in an attempt to spray the P-38. One of his guns still operated so Shipley fired to deter the enemy. When he attacked two other Zeros, and all of his guns quit, Shipley decided to practice discretion and dived out of the battle toward Nassau Bay.

Lt. Baker was leading the second element of Sparks' flight when they ran into a formation of eighteen Oscars. Sparks had taken the flight to Salamaua when John Lane had not received the message that about fifty Japanese aircraft were patrolling the Salamaua area. With Shipley on his wing, Sparks took after the Oscars when both formations had dropped tanks. Baker followed the lead. As the P-38s dived, Baker noticed an Oscar crossing to the right and got in a good deflection shot just as another Japanese got on his tail. Baker managed to loose the enemy in a dive and zoomed back up to 18,000 feet, just below the solid overcast.

He then led his flight toward two Oscars that were coming head-on from the right. One of the Japanese thought better of it and broke, but the other Oscar came straight on. It only took a few seconds for the gap to close and Baker got a good burst in before the enemy roared past. There was no time to check the damage, however, when two more Oscars came from another direction. Baker again turned into the enemy, but his flight had apparently lost him and the agile Oscars soon outmaneuvered his Lightning. With a hostile fighter behind him on either side, Baker shoved the throttles forward and sent the P-38 into a dive. Two frustrated Oscar pilots soon had to give up the chase. Once more the P-38 zoomed to 18,000 feet and Baker decided to return home since the battle had obviously dispersed.

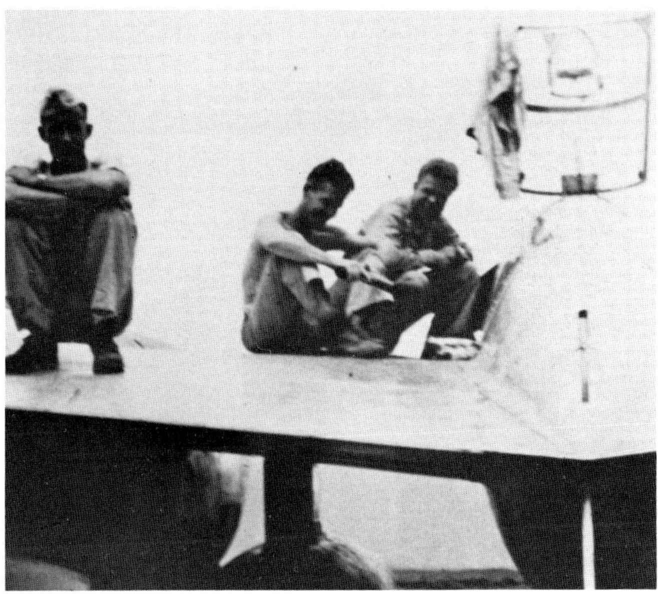

L. to R., H.L. Denton. T.J. Lynch and Dick Bong have a discussion while lounging on the wing of a P-38. J. Lane

Even with these combats taken out of context from the total picture of the battle, it was evident that the Japanese were again responding vigorously to the boldly ranging Fifth Air Force. During the next few weeks, Lae-Salamaua and Wewak would be threatened by Allied air and ground power. Incidently, during the combat of July 18, John Lane became the latest ace of the 39th when he downed one of the two Oscars claimed by the Squadron. More would follow in the days to come.

On July 21, 1943, the 39th passed a significant milestone in its history. On that day it claimed no less than eleven Japanese for a total of 104 confirmed air victories, making it the first Fifth Air Force squadron to attain that distinction. (The 9th Fighter Squadron would go over the one hundred mark on July 26 with 105 victories; the 80th Fighter Squadron would reach the century mark on September 26.)

Dick Smith claimed two Tonys and Paul Stanch claimed two Oscars to become aces on July 21, and Stan Andrews claimed a Tony to become an ace. As it was with most other American pilots at the time, Andrews was convinced that the Messerschmitt Me 109 was operating in the theatre. He was frustrated in trying to convince Intelligence of his conviction and wanted to see the gun camera film. Andrews was enlightened by the evidence of the film and believes that the first clear pictures of the Tony on the New Guinea front were made from his camera.

Other victors on July 21 were Ken Sparks, Denton, Duncan, Prentice (brother of the former commander of the 39th) and King, each with one Japanese fighter. For Sparks, it was his eleventh and final victory. He had been one of the wildest pilots in the Squadron, and his heavy toll of enemy aircraft was due partly to his unruliness. Sometimes he threw himself into enemy formations, often without the necessary item of a covering wingman. It is doubtless that he accounted for a number of Japanese aircraft, but throughout his career he claimed everything he shot at, diluting his actual successes.

Whatever his excesses may have been, he was certainly brave and eager for combat. His last report of July 21 demonstrates his fervor and abandon in battle:

"...My first pass was at one that came in from my right. I fired a short burst but no results observed. My second was a quarter right front attack and I fired about 7-9 seconds working around to quarter rear. Another plane attacked me but while evading the second I saw the Oscar I fired at roll about three times horizontally and then fall out of control. I had to employ active action in evasion, but turning about 10 seconds later I saw smoke coming from the area above the trees where the aircraft I shot had disappeared. At the same time I saw another plane spinning violently, as if it had a wing gone, catch fire, and crash into the trees about 2 miles from where mine crashed. I later found out that this second one to crash was shot down by Lt. Stanch. I made another pass at a Zero, but no results were observed."

Ken Sparks returned to the U.S. shortly after this and was killed in a flying accident in March 1944.

Paul Stanch was in the flight behind Sparks and attacked a number of Oscars, one of which may have been Sparks' assailant. Again the difference between the Nakajima Oscar and the Mitsubishi Zero is blurred in the mind of the American pilot:

"At point blank range (approximately 200 feet) I opened fire on an Oscar. I observed my bullets enter the engine. I brought my fire back to the cockpit and then out to left wing. The left wing of the Oscar sheared off about one foot out from the fuselage. The plane went into a violent spin immediately and went straight in, trailing black smoke. I pulled up in a climbing turn to the right and saw my wingman put a good burst into an Oscar and left it trailing black smoke. I saw a Zero on the tail of a P-38 and I made a diving 90 degree attack. I saw my tracers rake through the Zero (100 yards) but did not observe any results. A second later I saw three Zero's diving on another P-38. I made a diving tail attack on these three Zeros and opened fire at about 2,000 yds. I raked my fire from one Zero to the other but did not seem to get any results. I then centered my fire on one of the three zeros and held a three-second burst.

"Smoke started to pour out of the engine cowling and completely covered the cockpit of the Zero from view. I observed the Zero going in, in a slow diving turn to the left. I got in two 45 degree deflection shots a little later on, but did not observe any results. I left the area at 1230. I was out of ammunition. I saw 3 Zeros off to the side of nine B-25's that were headed for home. I made a climbing rear approach on the Zeros and as soon as they saw me they left the B-25's and started for me. I immediately started a shallow dive away from the B-25's. The Zeros followed me until I saw that the B-25's were out of sight, then I gave it full throttle and walked right away from the Zeros. I observed Lt. Denton shoot a Zero down in flames. I also saw Lt. Duncan destroy a Zero, and Lt. Smith destory a Me 109 type.

"I claim 2 Zeros (Oscars) certain and one damaged."

J. Lane

John Rogers flew the "Jolly Roger." The sharkmouth was a distinctive design of the 39th F.S. in the Pacific area of combat.

R.M. Hill

R.M. Hill

Above; Curran "Jack" Jones' P-38F No. 20, left; line of "Cobra" Lightnings at Jackson airfield, Port Moresby, Nov. '42.

It was a hot, uncomfortable job keeping the planes ready for combat and the 39th's ground crew had an exceptional record of having the fighters available for any and all missions.

Two days later, July 23, the Squadron escorted transports to the same general area of Bogadjim and ran into another big fight between Wau and Lae. Again the 39th claimed five Japanese fighters and suffered no losses. From the combat reports of 39th pilots it is obvious that the battles were not one-sided affairs. Combat for July 1943 was against skilled but desperate Japanese defenders. Few replacements were able to trickle in for the benefit of Japanese crews, and they fought with commensurate ferocity.

The edge for the 39th was psychological. The pilots believed in themselves and the aircraft they flew. Their battles were now on the offensive, and they patrolled enemy territory with resolve. Charles King kept a personal record and his entry for July 23 exemplifies the 39th's aggressive spirit:

> "Leading Sqdn. on top cover for Transports Wau-Bululo area. Given plot just after going above partial overcast after the first transports left. Saw P-38's (9th Sqdn.) and B-24's. Then saw 16-20 enemy Oscars and Type III's. At the same altitude coming from over Lae. Attacked. They stayed in a nice bunch and in 4 ship flights scissoring so they would catch (the) tail of your flight (which was) not with you. This happened after nil result (in) diving pass to the right. Dove down the coast and shook him and came up over Salamaua. Got wingman and 2 or 3 others and dove from 20,000' on Zeros at 14,000'. As I reached 16,000' observed 16-20 more coming in over Lae, at 1820 (18 to 20,000 feet). Was able to pull up into low flight only. Nil results on slight diving pass to the right. Made a dive out to left with two following, one firing at a fair distance. Dove to 12,000' with him following pretty far back. Just as he gave up my left engine started to smoke and I saw a hole outside of engine just in front of the supercharger. In 1½ minutes the Prestone (temp.) started up so I feathered the prop. As I passed Bululo saw the transports and P-39's also 1 P-38.
>
> "Hell of a time getting wheels down. (used both systems and landed at Jackson, 7-Mile). Plane complete loss due to burnt left boom."

Paul Stanch also scored during the same mission. His flight spotted a formation of about twenty enemy fighters which immediately went into several ragged Lufberry circles. Stanch quickly positioned himself above and attacked:

> "My first pass was a diving pass through the Lufberry circle. I fired a three second burst but did not observe any results. My second attack was a quarter tail attack. I fired

C. King

Major Charles King by one of two P-38s he normally flew.

a three second burst and I saw my shots enter the left side of the fuselage and left wing of the plane. The enemy aircraft went into a spin to the left. I did not observe any fire while the plane was going down, but there was a large flash when the plane hit the ground. The plane crashed about ½ way between Lae and Salamaua and about one mile inland.

"I saw a plane crash that I later learned had been attacked by Lt. Baker. This plane crashed between Lae and Salamaua about five miles inland. I believe it was an Oscar.

"My third pass was a head-on pass with an Me 109 (Tony). I saw tracers enter the cowling but did not observe results.

"I had one diving attack on two Zeros that were flying close formation over Lae. I sprayed about a four second burst back and forth through both of them. I did not observe any results because my plane started to buffet and I had a hard time pulling out of the dive."

Total claims for the fighters of the Fifth Air Force in July came to seventy-seven Japanese aircraft. Of these, sixty-three were granted to pilots of the three P-38 squadrons. The 9th Fighter Squadron was just beginning its impressive record and claimed twenty-two victories during the month. Another twenty-two were claimed by the 80th Fighter Squadron which had started operations in April with the P-38. In August the first complete P-38 Fighter Group, the 475th, was destined to begin combat. The 39th had set an example that was blossoming into a formidable aerial offensive.

By the time August came along, the 39th had been based at 14-Mile Drome of Port Moresby for over eight months, and some of the pleasures of rear area life were appearing. A new mess and recreational building, regular movies and other luxuries made the usually rugged Fifth Air Force base more like civilization. Considering the military propensity for removing troops from pleasurable situations, the 39th personnel must have looked nervously on the improvements as an omen.

The 39th stayed at Port Moresby until it got into another big scrap on August 20. It was one of the first assaults on the Japanese base at Wewak, and Japanese twin-engine fighters were encountered by Fifth Fighter Command for the first time, also. Tom Lynch claimed two of the Ki-45 Nick types and reported his combat:

" . . . A twin engine fighter came in from the right and I turned into him and followed him in a right turn. I observed hits on his fuselage as I broke away. I then made a direct tail attack on him. I shot out his right engine then hit his left engine causing his left wheel to come down. Part of his tail blew away then a large object came out of the front. I believe it was the pilot bailing out. He passed directly over my head. The plane fell off to the right and went straight into the ground.

"I then rejoined the bombers. When we were about 60 miles away from the target, I saw a single ship pulling up along the right of the bombers about 2 miles away. As I approached him he turned into me. I got a quick shot and observed my hits on his fuselage and right engine. The right engine quit and the ship skidded back and forth then turned over on its back and dropped straight into the ground. We rejoined the bombers and came home."

"Shady" Lane was flying in Lynch's flight that day and observed his double victory. Lane reported seeing one B-24 blow up when it was hit by flak over the target. A few minutes later he and Lynch chased a Zero that appeared uneager to confront them. Subsequently, Lynch got his two victories, and Lane saw the first Nick crash and the second explode. These two victories represented Lynch's twelfth and thirteenth and brought him close to top-ranking Dick Bong's with sixteen.

The next day Lynch came even closer when he claimed his fourteenth victory off shore, near Dagua Strip. He dropped his tanks when his flight spotted four Oscars and made a head-on pass from below on one of the enemy planes. Pieces came off the engine of the little Japanese fighter as Lynch's P-38 flashed by. Lynch swung around and attacked the Oscar once more while it raced for cloud cover. The Lightning sent out one long burst from about 200 yards, and the enemy fighter disintegrated in smoke and flame just as it reached the cloud. Lynch then went after two other fighters but they managed to get away.

John Lane claimed his last confirmed victory on August 29 during another B-24 escort to Wewak. Again there were sixteen P-38s from the 39th at about 23,000 feet when the Japanese interceptors appeared. Lane recorded the mission in his combat report:

"At approximately 1100 hours we were escorting 6 squadrons of B-24's. We were the high cover. At this time Major Lynch called in a Zero at 3 o'clock. We continued on with the bombers, but the low cover was not in sight so we dropped down to close cover. As we were heading down I noticed one Zero make a pass on the bombers, the tail gunner returned fire. No results observed.

"About this time directly below us were approximately 10 Zeros—Oscars I believe. These began climbing to attack the bombers. We then made our first pass which was directly over Boram strip. Major Lynch took shots at one which split S'ed away. Vining attacked another at 10 o'clock.

"I was about 6,000 feet behind when the Oscar pulled up as if to attack Major. I then had direct tail attack. I poured a fairly long burst into him and he immediately blew-up. After that I was partially blinded by oil which was all over my windshield. This oil came from the Zero which blew-up. I then tacked onto Major Lynch and then made another attack. I had one Oscar turn in front of me. I poured another and this one flew through the fire. This Oscar was damaged as I observed my fire entering his wing.

"After this attack I turned and headed for home. I tried another but could not see my flight or the Zeros. What you can't see, you can't shoot at. My plane was slightly damaged. One 20mm hit my blast tube and exploded. The shrapnel made several holes in my fuselage and canopy, also camera damaged by Zero pieces."

There is no doubt that many experienced Japanese pilots remained, but the ranks of the fallen were filled by untried neophytes and a certain desperation attended their tactics. Often in 39th records for the second half of 1943 Japanese pilots displayed a reluctance for aggressive action. More often the enemy pilots would turn away from an attack or at least hesitate long enough for the P-38 pilots to take the initiative. Whole Japanese air units were destroyed in the New Guinea fighting and nothing prevailed to absorb the momentum generated by the Allied offensive.

September ushered in another period of victories for the Fifth Air Force after Wewak was written off as a danger to the Allies. On September 4 the 39th was engaged in covering the Australian 9th Division during a landing near Lae.

During the mission, Charles King ran into some mechanical trouble when one of the propellers on his P-38 went out of control. King flew his airplane around the base to burn off fuel when the RPM dropped and he radioed for an emergency landing.

What followed was embarrassing at the moment but became an amusing anecdote in retrospect for King and for the commander of the 35th Fighter Group. King told the tower that his "prop had dropped off", which was a common phrase to denote loss of power in electric propellers. The Group Commander was a P-39 pilot, which utilized a standard propeller, and scrambled to the strip to see a P-38 land with only one propeller.

When the Commander did finally talk to King he was upset that he used such confusing teriminology. King was upset that the commander should be so unfamiliar with his choice of phrases. Later, King thought it was funny that a commander and high-ranking Squadron member should become so excited about their narrow views of jargon.

Tom Lynch became more justifiably unhappy about the fact that the 39th was scheduled to cover B-17s that carried Generals wanting to observe first hand the action about Lae-Salamaua. Lynch, among others, thought that all possible protection should be scheduled for the impending paratrooper operations.

The 80th Fighter Squadron was experiencing a period of outstanding victories and had claimed eleven aircraft on the fourth. Other squadrons ran into Japanese opposition and claimed another seven aircraft. There was little opposition encountered by the 39th which accounted for two of the final tally.

Eighty-two transports were escorted by the 39th on the paratroop drop at Nadzab the next day, completing a pincers move against Lae.

For the next two weeks the bases around Wewak would be hammered by B-24 raids. Marilinin, But-Dagua, Morobe and Bogadjim were visitied by P-38 escorted bombers and strafers. Japanese resistence was spirited even if its potential was waning. Tom Lynch scored his sixteenth and final victory on September 16. He would be sent home for a brief stay before returning to Fifth Fighter Command and scoring four more times before his tragic death in combat.

September 22 saw the 39th covering the convoy that was landing troops at Finschhafen. The Squadron got into a big fight with a group of lathered Japanese fighters. Dick Smith remembers attacking a black-painted Zero and sending it into the sea for his seventh and final victory. Six other Japanese aircraft were credited to 39th pilots, including one which collided with the P-38 piloted by Joseph Forest. Both planes were observed to crash into the sea.

Favorite fighter of the 39th F.S. . .the Zero Zilcher Lockheed P-38 Lightning. R.M. Hill

Losses had been amazingly light in the Squadron considering the rough missions that had been flown since operations began in May 1942. P-38 operations especially produced a high victory-to-loss ratio. By the end of September only about seven P-38s had been lost in combat circumstances against one hundred and thirteen victory claims. Lee Haigler, for example, had been reported missing in action during a mission over Lae in which no enemy were encountered.

Charlie Sullivan had lost his P-38 during the mission of September 18. He crash landed in the Ramu-Sepic Valley and walked back through some of the world's worst terrain for more than two weeks. At one point he was forced to shoot two hostile natives who apparently wanted to deliver his body to the Japanese.

With the neutralization of Wewak and the occupation of Lae the Japanese defenses were forced far back. Hollandia, Alexishafen and other areas still held by the Japanese would come under the same pressure as Lae and Wewak. There was no reason for the enemy to believe the result would be any different. Whatever respite the defenders may have enjoyed was momentary and presaged the tide beginning to sweep the northwestern coast of New Guinea.

There was only one thorn in the side of the Fifth Air Force that had to be removed before the procession could begin again. That thorn was the airpower of the great bastion which posed a threat to every Allied move in the Southwest Pacific.

Rabaul, Nov. '43

On the northeastern side of the island of New Britain is a sprawling harbor which borders the town of Rabaul. When the Japanese occupied the area early in 1942, they set about making Rabaul a key base from which several potential targets, including New Guinea and the Solomons, could be reached.

C. King

Kawasaki (Lily or Nick) downed by Tom Lynch. . .from his gun camera film. Victory claim of Sept. 4, 1943.

Four Major airfields existed around Rabaul's Simpson harbor (Lakunai, Uunakanau, Rapopo and Tobera), and many of the crack fighter units, such as the formidable Lae Wing, were moved into the base when American attacks became imminent during the latter months of 1943. In October the Japanese anticipation was realized.

Charles King had taken command of the 39th when Tom Lynch went home in September. Fighter Command sent King up to the island of Kiriwina, midway between New Guinea and New Britain, to check the facilities built by Allied engineers for the servicing of P-38s. King made the trip on October 7 and noticed, among other things, that the flight to Kiriwina took an hour and thirty minutes and the installations on the island were quite suitable for damaged or malfunctioning fighters. One drawback of Kiriwina, however, was the lack of radio aids. Navigational guides were to be provided by bombers flying near the island.

Another problem that was to be forbidding as the enemy was the weather. It was not unusual for violent fronts to suddenly build up for thousands of feet, threatening to force smaller single-seat aircraft off course, or worse . . . cause loss of control. For the eight successful raids flown against Rabaul between October 12 and November 7, at least ten were cancelled by weather.

Bombing raids against Rabaul had been carried out since the beginning of Fifth Air Force operations, usually at night and never before with fighter escort. On October 12 more than two hundred bombers escorted by one hundred and twenty-five P-38s attacked Rabaul and claimed more than twelve fighters in the air and many others on the ground. After the mission of the 12th some P-38s were used on sweeps during the raids and large numbers of Japanese fighters would be claimed in the free-swinging slugfests that developed.

One such fight took place during the next raid on October 23 when many more defending interceptors were claimed. Weather had interfered with planned strikes before the 23rd but on that day forty-seven P-38s herded the bombers over Rapopo airfield. Two P-38s were reported missing when the mission ended.

Several Zeros tried to interfere with the 39th while it flew cover. Paul Stanch scored his tenth and final victory in spite of a rough situation when his airplane developed mechanical trouble:

"On approaching target (3 ship flight) I saw one Zero (Oscar) 10 o'clock and same level. I did not press attack, but called him into the Squadron. I saw 10 to 12 Zeros flying among the Ack Ack burst and they were very hard to observe. They were about 2 or 3,000 feet above our level at 3 o'clock. They started to dive towards the bombers and I turned my flight into them for attack.

"My right engine started to stream prestone and overheat. I pulled under the Zeros and 6 of them split-S'ed down and the rest chandelled to the left. The 6 Zeros did not get in range, but they followed my flight and I led them away from the bombers. I could not draw enough power to lose them, but they did not gain on me.

"One Zero came in high 12 o'clock and I started to climb into him. He did several vertical rolls. The Zeros behind were gaining on me so I leveled out. I saw another Zero 1 o'clock and same level. I turned into a head-on pass. I fired from 500 yards to 50 yards. I saw my bullets explode from the rear of his cockpit to the tail as the Zero started to roll under me. He passed about 20 feet below me, and that was the last I saw of him. S/Sgt. John W. Price, a gunner in the bomber formation saw this plane hit the water and explode. I went on back to the bombers and stayed with them for about 20 minutes. I then went straight to Kiriwina and landed at 1420/L."

General Whitehead decorating Maj. Frank Royal, Jan. 15, 1943.

For the next two days Rabaul was attacked by hundreds of B-24s and B-25s. Fifth Air Force records claim that about fifty Japanese fighters fell to the P-38 escort between October 23 and 25. The 39th was credited with five victories for the three day period.

Aerial combat over the stronghold was a vicious affair. Neither side gave quarter. One excerpt from the combat report of Lt. Lloyd Shipley for October 24 displays the kind of casual brutality:

"We had flown as far south as Wide Bay when the B-25's reported an enemy at 3 o'clock. I saw him and turned toward him, whereupon he made a run for home. I took up the chase and by this time I had five P-38's following me. At 40 inches and 2600 RPM I easily pulled up into range and opened fire. I saw one of my explosives hit near his wing tip and he was leaving a faint trail of smoke, which was undoubtedly (sic) due to wide open engine operations. At this inopportune moment my guns gave out of ammunition so I pulled up to give the men behind me a clear field. Lt. Walters shot up the Zero so badly that the pilot bailed out.

"The plane continued on its dive into the water. The pilot was strafed and killed."

An exception to the general uncompromising mood of the battles occurred in the case of the only 39th casualty during the campaign. The Squadron usually sent new men out on rela-

Australian " Boomerang" fighter taxiis out for a takeoff after a visit to the 39th base area. J. Lane

Major Lynch, Lt. Col. Prentice and 35th Group Commander Col. Legg at the party celebrating the 39th squadron's 100th aerial victory.

S. Andrews

tively safe missions for their first sorties. In the case of Lt. Alfonse Quinones, however, the first mission was scheduled during the Rabaul operations.

Quinones was assigned to what was considered a safer part of a fighter formation; the number two man in either the second or third flight. On one of the November missions, a sharp-eyed Japanese fighter pilot fired a deflection shot into a 39th formation and shot Quinones out of the middle. Crews shook their heads in disbelief and his chances of surviving the crash.

But King was surprised when after the war he had the pleasure of answering an Army request for evaluation on Quinones to accompany his application for a regular commission. It seems that he had managed to escape his doomed fighter and was picked up the moment he landed by the Japanese who inturned him as a prisoner of war.

On the 29th of October, by the way, King managed his last two victories over Rabaul to become the final P-38 ace of the 39th. During this particular mission he utilized an experimental installation of color film in his gun camera. Whether or not that fact helped his fighting efficiency is not as important as the film's clarity when it was developed; every detail of the two Oscars and their demise was shown. King was promised a copy of the film but the original is undoubtedly still floating around the Pentagon awaiting final evaluation.

One thing that surprised the Fifth Air Force during the Rabaul operations was the fact that the enemy's strength in the air seemed to grow rather than diminish in spite of the losses sustained. Some Navy carrier units flew in their Zeros in a desperate attempt to bolster the island defenses. Postwar research into Japanese records is frustrating at best since exact strength of forces at Rabaul and rate of attrition is not clearly defined. The effect of the Fifth Air Force campaign is therefore difficult to determine.

Whatever the Fifth Air Force contribution, nevertheless, Rabaul did succumb to the heavy pressure. During November and December the Thirteenth Air Force and Navy and Marine air units began to strike at the Japanese base. By March 1944 all Japanese air units had been evacuated. The Fifth Air Force was free to continue to march along the New Guinea front.

The last victory scored in P-38 aircraft by the 39th Fighter Squadron came on the Rabaul mission of November 2. Hamilton Salmon teamed up with Lt. Allen Hill of the 80th Fighter Squadron to do a job on the Zeros which attempted to attack some B-25s. Hill downed at least two Zeros to become an ace and witnessed Salmon's claim. The two P-38 pilots lined up behind a pair of Zeros and Hill fired a three-second burst at the one on the right to send it down in flames. He then looked around and saw Salmon's quarry smoking and falling out of control.

At the end of the Rabaul operation many of the older hands in the Squadron were sent home on rotation. Charles King remained in command of the 39th until he turned the position over to Harris Denton in the middle of December. Stan Andrews went home in December and returned to the Squadron briefly in 1945. John Lane and Dick Smith stayed on through the spring of 1944. Curran Jones had left the unit in May 1943 to become an instructor.

Another old hand would be retained to take command in the middle of 1944 . . . Major Richard T. Cella. He had as much effect on the 39th as Tom Lynch or George Prentice. Cella was one of the first P-38 pilots to reach the 39th and was influential in developing confidence with the aircraft. Cella also helped improve the quality of gun-camera film by replacing the vibration-prone nose mounting with the starboard wing shackle. Cella led the 39th until the fall 1944.

Charles King left the Squadron in December but managed to remain in command throughout the conversion to P-47 aircraft. Pressure from the Japanese had been generally dissipated so the change over took place at a more relaxed and orderly pace than did the P-38 change. King flew one of the P-38s back to Dobodura on November 19 and two days later started to check out in the P-47D.

About the only unhappy note King could remember about the P-47 was its limited range. The airplanes on which the 39th converted were beat-up clunkers, and King had not flown anything but tricycle-gear aircraft since flight school. But he, like most others of the Squadron, took the change dutifully. In any event, he was surprised by orders waiting for him at Nadzab to return home and by December 22 he was at Brisbane looking forward to a flight home.

With the change to the P-47 there would be less aerial combat for the 39th. The Squadron had ended its year of the whirlwind and would enter a year of the doldrum. Another twelve months would pass before extended victories would fall to the unit. In that time other units flying the Lightning fighter would far surpass the pioneer 39th. The 39th would enter a period of yeoman service in cover and sweep operations. Its days of glory were over, but it would do the work of the plodding warhorse with equal devotion.

R.M. Hill

New equipment came in the form of the big, powerful P-47 Thunderbolt, a true fighter bomber. It ushered in a new dual role for the 39th.

Operations with the P-47

The beginning of 1943 brought only a few but nevertheless significant contacts with the enemy. Perhaps the only serious complaint Fifth Fighter Command pilots had with the P-47 initially was its range. It could not reach the distant areas where Japanese airpower had been forced. After Cape Gloucester and Finschhafen were secured, the nearest Japanese opposition was in the Hollandia region.

One promising aspect of the conversion to P-47s for the 39th was that they had the advantage of experienced instruction. The pressure under which the P–38 was introduced to combat no longer existed since the Allies were on the offensive in New Guinea. By the time the 39th had moved into Nadzab its pilots were experienced in P–47 operations and had the advice of Thunderbolt pilots with combat time to avoid the mistakes committed initially with P–38s over Lae.

Since the role of the entire 35th Fighter Group changed from defense to support of the Sixth Army in its march from the Huon Gulf to Hollandia, the choice of the P–47 was especially fortuitous. Long before January 1944 the P–47 had earned a formidable reputation for ruggedness and deadly effectiveness. Fifth Air Force reaction to the fighter was lukewarm at best and some commanders were outright abusive in criticism. But the P–47 found its place as ground support for the Allies in New Guinea and proved to be an effective bomber destroyer, especially at high altitude where the earlier fighters were rather defficient.

The first P–47 operations undertaken by the squadron came as soon as the 39th moved up to Nadzab on December 15, 1943. For the first time since they entered combat the three squadrons of the 35th operated as a unit from the same location. Much of the tedious patrol activity conducted by the 40th and 41st was relieved by the appearance of the 39th.

Perhaps the first important contribution that the 39th made to the New Guinea campaign with the P–47 came on December 25–26 when the squadron covered the amphibious landings at Cape Gloucester and Borgen Bay. Although no enemy air opposition was met, the 39th supported the landings which met with bitter resistance on the beaches. After Cape Gloucester was secured there was only scattered Japanese activity on New Britain and the Solomons campaign was coming to an end.

During the month of December only forty-seven missions of all types were flown, culminating in the Cape Gloucester operations. A period of heavy operations was coming for the squadron but very little of it would be the more glamorous air-to-air fighting that had made the 39th Fighter Squadron famous during the previous year.

Nadzab to Gusap, Jan.-Mar. '44

Nadzab was a pleasure for 39th personnel considering how other operational bases were constructed. Some planning went into the site location and building plan. Not only was the base relatively free of the usual New Guinea pests but facilities were conveniently placed and the crews enjoyed fresh water with adquate protection from the dust and rain that alternately plagued that part of New Guinea. There was even some consideration given to the human need of relaxation by the presence of recreation halls.

During January the training on the P–47 continued while new pilots were added to replace those rotated home. On January 3 and 4 a total of eight pilots signed into the squadron. 2nd Lieutenants Maurice Boland, Telesphore Graber, John Lockhart and Billy Richards came to the 39th after a tour in P–39s with the other squadrons of the 35th Fighter Group. 2nd Lieutenants James Steele, Robert Thorpe, Fredrick Tobi and Marcus Trout were brand new to the theatre but had some previous time in P–47s.

Troop landings were made at Saidor on January 2, which the squadron covered. The January 8 operations gave the 39th its only other action during the month. Very few air victories were scored during 1944 but the first came during the January 8 mission as is described in the narrative combat report:

"The 16 planes tookoff at 1000/L,from Nadzab to make a fighter sweep over Wewak. 13 P-47's arrived over Wewak at 1120/1, altitude 23,000 feet.

"The formation made two circles over the area underneath an overcast at approximately 23,000 feet. At. 1145/L, enemy planes were sighted. Captain Denton, the squadron leader, started to turn into them and climb up to their altitude. Another group of planes came through the overcast and attacked the formation from the rear and above. The formation went into a steep dive to evade the enemy planes.

"Col. Doss and Lt. Tansing could not drop their belly

tanks, so they truned to the right and left, respectively in their dive. One of the Zekes got on Col. Doss's tail. Lt. Clymer was able to turn and get in a full deflection shot on this Zeke. It went into a diving turn to the right, started smoking, and burst into flames. This Zeke was destroyed. As Lt. Tansing was diving to the left, a Tony passed in front of him and he got in a snap shot with nil results observed.

"Approximately 30 Zekes and Tonys were sighted flying four ship flights and 2 ship elements. All of the planes observed were a light grey or silver with red roundels on the wings.

". . .The pilots opinions were that the Japs knew they were coming and what they were. Four large vessels in the area took no evasive action as they would if bombers were coming, and it seemed the fighters had been sitting on the overcast waiting to attack when they knew the planes could not stay much longer. The Ack Ack did not open up until the belly tanks were dropped. Several bursts were observed around 22,000 feet which were accurate as to altitude."

As it happened, Lt. Clymer's victory was never confirmed, leaving the entire 35th Fighter Group scoreless in January 1944. Support missions became the routine and transition to the P–47 was still taking time. There were some items like fuel systems to which the pilots had to become accustomed, and the Thunderbolt had some idiosyncrasies of its own. Pilots, used to the torqueless P–38 or gentle P–39, were sometimes rudely surprised when a P–47, thrown carelessly into a turn, would begin autorotation around its own engine. The pilot had to take great care in coaxing the P–47 into a turn and it took some time to learn that trait.

On the fifteenth of January, Japanese aircraft paid a "Welcome-to-Nadzab" visit and sent the base personnel scurrying for cover. Eight Tonys strafed the field and destroyed one P–47, damaged four others and wounded Sgt. Charles Collins slightly as he ran for a slit trench. When the 39th people reached safety they gazed up in wonder at the fireworks going on around them. The Tonys raced from one end of the field to the other in a brash display of military fortisimo then got away clean after doing considerable damage.

Lt. Lloyd Shipley was one of the old P–38 pilots who transitioned to the Thunderbolt and he ran into some trouble with his plane on January 16, the day after the Japanese strafed Nadzab. Shipley was taking off on a morning patrol when, just at the critical moment of leaving the ground, the P–47 developed a runaway propeller. He cut back on the power and was about to make a successful abort when some tar barrels loomed ahead and blocked his path. The big Thunderbolt crunched into the stack of barrels and effectively tore them and itself to pieces. Shipley, however, protected by his sturdy mount, walked gratefully away from the wreckage without harm.

Morale amongst enlisted men was noticeably sagging until a rotation policy came into effect in January for officer's candidate school and should be mentioned as a symbol of the non-commissioned effort in the record of the 39th. MSgt. Celeste Martini had worked on the armament problems experienced with the early P–38s. He devised a bracket which eliminated the damaging vibration of bucking .50 calibre guns and worked out the stoppage problem that limited the efficiency of the P–38s massed firepower. His achievements were typical of many behind-the-scenes activities that helped the 39th improve their skills.

It also became evident that Nadzab was unsuitable for both fighter and bomber traffic which the base had to accommodate. Shortly thereafter an order was given for the 35th Fighter Group to move to Gusap. The ground war was moving forward and air power had to move with it.

By the first part of February, ground elements had prepared a site at Gusap, located in the Ramu Valley, northwest of Lae. The 49th Fighter Group had moved into the area some weeks before the 35th Fighter Group arrived and personnel of the 49th did not seem to get a very positive image of the flying ability displayed by its sister unit. Some members of the 9th Fighter Squadron reported that whenever a P–47 of the 35th made a takeoff, other personnel on the five Gusap airfields would give it a wide berth.

Apparently there were a number of accidents happening to the 35th at this time. One 9th Squadrons pilot typified the situation with, "That bunch was born to trouble. . .". If, in fact, the P–47s of the 35th were tripping over their own feet, it does not seem that the 39th Fighter Squadron was participating to any large degree. There isn't any special mention of accidents or flying problems in 39th records for the period.

There was a paratrooper scare on February 9 that had the edgy Gusap troops even more spooked than before. What with the threat of strafing or their own P–47s falling on their heads, Japanese paratroopers were too much for the frayed nerves of the crews. The rumor became just that; somebody's overinflated interpretation of a report that low-flying enemy planes were seen in the area. All day long personnel at the Gusap bases sweated in hurriedly prepared bunkers waiting for the first Banzai. Late in the day the defenders of the mini-Almos cautiously came out of the security positions and shrugged off their reaction to the phanton threat.

Missions during February were accomplished in spite of the hazards involved. For the most part, the 39th made the usual patrols and sweeps over the Wewak and Hansa Bay areas. On February 12, another old hand in the squadron moved on when Capt. Paul Stanch was transfered to 35th headquarters to become assistant Group operations officer. The nature the 39th was changing slowly as the war in New Guinea progressed.

March interlude

Throughout March, April and May 1944 the 39th participated in patrol, bomber-escort, strafing and dive-bombing missions, mostly in the Wewak area. One such mission was an escort of B–24 bombers. Lt. John "Shady" Lane was heading the 39th formation in company with the group commander, Col. Furlo Wagner. Seventeen P–47s of the squadron circled over the target for ten minutes while the Liberators made their bomb runs. The fighter pilots were frustrated when Japanese interceptors appeared, took a quick look and decided to go home.

While this particular mission was taking an uneventful course, four more upstart Kawasaki Tonys slipped into the Gusap area and roared in to strafe number-5 fighter strip. As fortune would have it, Major William McDonough of the 41st Fighter Squadron was just returning from an aborted mission when one of the Tonys snarled directly across his landing approach. McDonough was rattled for a moment, then retracted his landing gear and slipped onto the tail of the intruder. After one short burst from his guns, the Japanese reared up and nosed over to crash into the ground before dozens of witnesses on the ground. It was one of those days when the squadron would have done better if they had stayed home.

On the next day, two former members of the 39th, who

were now flying P—38s directly for V Fighter Command, flew a sweep over Wewak area. Tommy Lynch and Dick Bong had become close friends since their days with the 39th. Lynch was able to check the wild-flying streak in Bong. Bong in return had a great admiration for Tommy as a person and in his ability as a combat leader.

The two pilots were at 17,000 feet over the target area around 1:30 in the afternoon when they sighted five Oscars flying below toward Dagua. Lynch and Bong dropped their external fuel tanks and dove toward the rear of the enemy formation. Lynch's combat report then describes his twentieth and final confirmed victory:

> ". . .(I) made a tail attack on the rear plane in the enemy formation. The Oscar burst into flames and fell off flaming from my first burst. I made several more passes with nil results and then sighted 6 more enemy aircraft approaching. I called Captain Bong to break off and we headed for the overcast at 6,000 feet. We started to return to base and then decided to attack again. I made a headon pass at an OSCAR and observed hits around the engine. A fairly large piece of metal came off his plane and hit behind my left prop, denting the engine cowling.
>
> "I made two more passes and then called to Captain Bong to break off. We returned to base flying straight down the coast over the water."

It was only four days later that Lynch flew his final mission. He and Bong were flying their P—38s over the Tadji area looking for the enemy when they sighted some luggers in the water below. The pair went down in what was meant to be a single pass on the ships. Lynch's plane roared over the target at about 300 miles an hour. As he was pulling out of the dive one of the propellers wrenched loose and flew off. He tried to make shore and managed to bail out just as his fighter exploded.

Bong spent a futile few minutes circling the area looking for his partner. "It cost me one of my best friends", Bong later recounted, "and it cost the Air Force one of its best combat pilots". Bong and Lynch had claimed seven Japanese aircraft between them during the few weeks that they flew together but now the peerless former commander of the 39th Fighter Squadron was dead.

Another loss befell the 39th a few days later, on March 14, when 1st Lieutenant Gene Duncan was returning from an escort to Wewak. His engine cut out and he was forced down into the sea near the mouth of the Sepic River. Since there were no rescue facilities at that time, other members of the squadron covered Duncan until they were forced to leave due to lack of fuel. Only Lieutenents Leroy Grosshuesch and James Querns continued to hover over the downed pilot until darkness fell. Although many rescue attempts were made, Duncan was never found.

The Japanese ventured another bombing of Gusap on March 17, when a lone enemy plane slipped into the area during the early morning hours and dropped three bombs. Although the crews were shaken by the wily foe's ability to sneak past the warning network, no particular harm was done. After that, the rest of March passed peacefully. Most of the air activity was taking place around Hollandia and very little was left for the 39th to participate in.

Staging for the Philippines, Apr.-Sept. '44

Although Hollandia was outside the range of Fifth Air Force fighters, General Kenney had received fifty-eight P—38s in February with greater fuel capacity, making flights into the area possible. Along with other modified P—38s, the ability to strike directly at enemy concentrations around Hollandia was at hand. Between March 30 and April 12, 1944, a series of raids were carried out which effectively reduced air power in the region. At the same time the 39th escorted missions to Hollandia, Wewak, Tadji and Hansa Bay. They also strafed targets of opportunity upon returning from these missions.

P-47s of the 35th F.G. sported theatre area white tail surfaces with their own pre-war rudder stripes and individual squadron engine cowl blaze. R.M. Hill

It was during one of these escorts that another loss was suffered by the squadron. First Lieutenant Albert H. Lane was preparing to take off on April 5 when, as his airplane left the ground something unexplained happened, and Lane crashed into the base operations building. The fighter exploded into flames, killing the pilot and an operator in the shack.

Amphibious landings were made at Hollandia and Aitape on April 16/22 and thereafter no major Japanese force was able to contest northern New Guinea. A few days earlier another 39th veteran, Dick Smith, ended his tour and went home on rotation after twenty months of Pacific combat.

The strafing and escort routine continued during May. By the 25th squadron personnel were surprised to receive orders sending them back to Lae where they had fought so many of their successful battles the previous year. When the fighting for Biak did not go as scheduled, the intended short stay at Lae lasted a full two months. It wasn't until June 19 that the first echelon of the 39th sailed from Lae, only to get as far as Hollandia where they were once again delayed, this time for almost a month. On July 20, the ground echelon sailed and arrived at Owi Island the next day. A temporary campsite was established on the eastern coast in anticipation of the air echelon's arrival.

Meanwhile, the pilots underwent training at Nadzab until the Owi base was ready. On June 25, seven new P-47D-23s arrived from Finschhafen. Seven more Thunderbolts were added two days later but no missions were flown until August 3, when glide-bombing was introduced against Japanese positions on Noemfoor Island and the Vogelkop peninsula. For the first time 500 pound bombs were used on the underwing shackles, doubling the P-47s offensive load.

During June another change is command took place when Major Harris Denton turned the squadron over to Major Richard Cella. Cella was one of the first P-38 pilots to reach the 39th in the summer of 1942. His contribution to the squadron's record was, like MSgt. Martini's, more technical than operational. Advice, such as Curran Jones had received to slam a malfunctioning P—38 back on the runway was countered by Cella's more knowledgeable instruction to cut power on the good engine and properly trim the airplane. Pilots developed greater confidence in the P—38 because of Cella. Another contribution that Major Cella made allowed the squadron to get reasonably clear gun-camera film such as Charles King's color pictures of his Rabaul victories on October 29, 1943. Cella took the camera from beneath the vibration-prone P—38 nose and replaced it under the starboard wing inboard.

The first mission commanded by Cella came on August 7 when he led a flight of twenty-six P—47s on a 900 mile trip from Nadzab to Noemfoor. The purpose of this and subsequent missions was to cover the building of airfields on Middleburg Island. For the next six weeks the 39th guarded against the occasional raiding of Japanese soldiers during the tenure on the island.

August 9 was the date of the first 39th operation from Noemfoor when two flights of P—47s covered a convoy off Cape Sansapor. Similar missions were conducted until August 20 when the squadron went out on a fighter sweep to the Moluccas, led by Captain Gordon Prentice. With a flight duration of five hours and twenty minutes, it was the longest P—47 mission flown by the 39th (and probably any other Fifth Fighter Command unit) up to that date. No longer could the skeptics berate the P—47s range. Later in the year, with three external tanks and a technique of controlling the curising speed with propeller pitch, as demonstrated by Charles Lindbergh, the squadron registered missions up to eight hours. The Thunderbolt did indeed have its big league boots.

Personnel losses in combat were always sad, bitter pills to swallow but when they occured during routine missions and not caused by enemy involvement, it was especially hard to take. Some pilots simply disappeared on routine missions. To a fighter pilot, the one on one encounter with an adversary was what he was trained for. There is something unexpected and seemingly unfair when he is downed by other means. Lt. Billy Richards was shot down by enemy anti-aircraft fire over Jefman Island. Richards had joined the squadron only the preceding January and was the first combat fatality suffered by the 39th over the northwestern New Guinea area.

Lt. Billy Richards, killed over Jefman Island during a dive bombing attack, Aug. 23, 1944.

On September 5, still operating under the 309th Bombardment Wing a Glide-bombing operation was carried out on Halmahera. With external tanks and a 500 pound bomb on each P—47, the 39th made the 800 mile round trip with no difficulty and all aircraft returned safely and even managed to have plenty of fuel to spare! Operations were curtailed briefly again when the air echelon began its move by air transport to Morotai Island on September 12. Four days later the ground contingent moved from Owi to Morotai on the transport, ***Andrew D. White.*** Combat operations began again from Owi, but consisted of the usual patrol routine.

Because of primitive and inadequate facilities on Morotai, ground support elements were forced to remain in Morotai's harbor until October 5. After that, the squadron set up temporary camp with the 40th and 41st Fighter Squadrons. The 39th found themselves only a few hundred yards behind the fighting front and operating out of a banana grove. Although the 39th missed the action in which the rest of the 35th Fighter Group participated on October 10 and 14 against Balikpapan, it would participate in operations over the Visayan and Molucca Islands, Sulu Archipelago, the Celebes and in the coming invasion of the Philippines.

End of the doldrum, Oct.-Nov. '44

On October 15 the ground echelon of the squadron moved into permanent installations in a coconut grove on Morotai. The site was paradigm of South Pacific air bases with the incongruous thrust of engine and propeller blade against the low skyline of palm trees and fleecy clouds between blue sea and air. The airstrip itself was a giant cross of crushed coral hacked out of the verdant landscape.

Morotai differed from the rolling terrain of undulating hills and thick undergrowth of small bushes and trees on Noemfoor. Morotai was flat and the final camp was quite comfortable, everyone enjoyed the luxury of unimpaired sea breezes whistling through the palms. P—47s of the 39th were often framed against billowing white clouds and powder-blue sky as they lifted their silver metal finish into the air. The classic image of the Pacific island airfield was well established by the last quarter of 1944.

It was not until October 23 that the 39th air echelon flew up from Owi to Morotai and the squadron was together in the same campsite for the first time since the last days at Gusap. Operations began the next day under control of the 310th Bombardment Wing. Not since the closing days at Rabaul had the 39th as its main task that of escorting bombers. The situation soon changed dramatically.

On November 6, 1944, the 39th ran into a whopper of a fight over the Philippines and scored its first confirmed victories in a year. Field Order Number 7, Squadron number 294, dated 6 November 1944 indicated that the 39th was operating fifteen P—47s on a fighter sweep to Cebu at about 14,000 feet when four to eight Oscars and Zeros were encountered over Negros Island at about 1040 in the morning.

Lt. Richard Ross was leading Red Flight and the entire 39th Squadron formation when he ran into a trio of Oscars which started rolling into a cloud cover before he could make a pass. Then Ross sighted another pair of Oscars flying 90 degrees away. He quickly called Red Flight to drop tanks and

The 35th Fighter Group out on a hunt. Early P-47s were olive-drab in color but by the summer of 1943 all squadrons attached to the 35th Fighter Group had natural metal finished Thunderbolts.

USAF

the other three flights to provide top cover. The enemy was apparently not too alert, for they swerved not a bit while Ross attacked. His report states, "I commenced firing at 400 yards, breaking off at 50 yards and observed hits on the right wing and fuselage". The attack caused the enemy fighter to smoke and the right landing gear to drop.

The other Oscar managed to turn on the P–47s and Ross saw the uncomfortable sight of tracers passing his left wing. With a full-power climb, he was able to loose not only the Oscar but also his entire flight. When he was unable to locate the rest of his squadron, and with fuel at the critical point, he headed for home.

Carl Rymer, leading Outcast White Flight, followed Red Flight in the attack and tried to drop his tanks but the right one refused to release. He was frustrated in attempts to shake the thing off, but spotted an Oscar below and decided to attack anyway. The Japanese plane was coming from beneath some clouds and was probably part of the same formation that Ross had engaged. He dived on the Oscar and followed it around a mountainside, closing in at about three hundred miles per hour. Just as he got in range, dead astern and at no more than two-hundred feet off the ground, the Japanese pilot did a half roll and a partial split-S, coming out below treetop level with another half roll. Rymer overshot the little fighter when it made a sharp turn to the left.

Lt. Perkins, who was Rymer's element leader, made a pass at the enemy plane but he and his wingman both missed when the Oscar made similar evasive maneuvers. Rymer made another run but cursed his right wing tank when it forced him into a skid and his burst of fire went wide. When he finally recovered, he noticed Perkins fire a deflection shot which caught the Oscar behind the cockpit before he again overshot the target. For the first time in the action, the Japanese pilot was in firing position and let a burst fly at Perkins. His wingman made a head-on pass to drive off the pesky enemy.

The climax came a moment later when the Oscar pulled away from Perkins' tail. Rymer put a quarter radii lead below the enemy fighter in anticipation of his usual roll and split-S maneuver. When Rymer pulled the trigger, the Japanese flew through the line of fire and burst into flames. The Oscar fell about two-hundred feet and burned until it hit the ground and exploded.

Lt. William Rogers was leading Blue Flight at the same time the three Oscars were sighted. He quickly got his flight above the enemy but had a time getting his claim for a probable, which is indicated in his combat report:

"We came on first enemy plane from left full deflection and I wasn't able to lead him. However after I broke away climbing toward the second enemy plane, my wingman, Lt. Foster, followed the first and gained a victory as confirmed by the third man in our flight, Lt. Boland.

"I climbed toward second enemy plane and was able to maneuver on his tail out of range at 23,000. He went into a shallow turn and did a quarter roll to the left. I was able to lead him and started firing through my quarter roll and opened fire slightly out of range at about 600 yards. I held the lead and closed slowly as I held the trigger down, observing some burst. I followed him down in this manner to about 10,000 feet in an almost vertical dive, closing to about 350 yards.

"At this point my controls were fouled by gas line to left external tank which didn't release with the tank. The plane shuddered violently and I had difficulty pulling out of the dive which I did at 5,000 feet. I didn't observe the Type O, single engine, single seated fighter (Zeke)* after 10,000. This (fighter plane) took very little in the way of evasive action except for diving and I had no difficulty following him through the quarter roll. I fired about a 15 second burst from 30 degrees to 0 degrees deflection. I claim one Type 0, single engine, single seated fighter probably destroyed."

However, the flight led by Lt. Marcus Trout spotted the same group of Oscars and was able to maneuver his group into position to where Lt. Ernest Johnson was able to down one. Trout made a 45 degree deflection run into a dive at another Oscar. The Japanese made a rolling maneuver which allowed Johnson to follow him. At about 9,000 feet, Johnson and Trout were both close on the enemy's tail. Lt. Trout and another man in the flight, Lt. Jim Crowford, both fired, hitting the enemy just as Johnson roared in at about 230 miles per hour, and also fired a burst at close range. The enemy plane burst into flame and rolled over, falling into Guimaras Strait on its back.

After the brief but ferocious air battle had subsided it was determined that the 39th had accounted for three Oscars confirmed and two probably destroyed. The 39th lost Lts. Avant and Steele to unknown causes.

---

*It was more likely that Rogers attacked another Oscar, although he was obviously well occupied at the time with matters other than aircraft identification. He was also certainly frustrated at not being able to consumate the victory.

As a matter of record, the Japanese were badly organized in the Philippines. Two great sea battles had been fought in the area to the Americans favor and they dominated the routes of supply and communication thereafter. Much of the equipment necessary for maintaining an effective air force was simply not available to the Japanese. In addition, good air intelligence was denied them and many of their fighter pilots would sortie out in the latest model aircraft to make the same tatical mistakes that had claimed so many of their comrades earlier.

The stage was now set for the last period of aerial combat to be experienced by the Fifth Air Force. On November 7, 1944, Major Cella turned the squadron over to Lt. Leroy V. Grosshuesch when he was ordered back to the Command and General Staff School at Fort Leavenworth, Kansas. Grosshuesch would lead the squadron through the last stages of its World War II service while a ragged Japanese war machine fell quickly to the Allied Juggernaut descending on the home islands.

By the end of November, V Fighter Command had claimed no fewer than three hundred enemy aircraft over the Philippines. The fighter units that had moved into the new bases on Leyte just after the invasion of October 20 were enjoying unprecedented good hunting. While based on Morotai, the 39th had to be satisfied with only sporadic contact with Japanese aircraft.

Moving to the Philippines, Nov.-Jan. '45

As a consequence of the air battles fought during November, Fifth Air Force fighters based in the Philippines enjoyed a decided supremacy over the hard-pressed Japanese. The 49th Fighter Group had converted to the P–38 and had a total of well over 500 air victories by the middle of the month. The 80th Fighter Squadron of the 8th Fighter Group had taken the lead as high-scoring squadron in SWPA with over two-hundred victories until the 9th Fighter Squadron took back the honors for keeps later in November.

High status in the aerial victory count had long ago been relinquished by the 39th. By the end of the war, at least four other squadrons of the Fifth Air Force would claim more aerial victories than the 39th. When 1944 came to a close, the 39th had some 150 claims. Still there was never any lessening of spirit. Lt. Charles Covey, for example, came to the squadron during December and by the following spring had scored three confirmed victories in the only air battle he was to engage in. Covey was convinced that the only thing that prevented him from becoming an ace was the lack of air-to-air combat the 39th experienced.

Whatever the fortunes of war dictated, the 39th finally did have contact with the enemy again in the latter days of November. On the 21st, Leroy Grosshuesch, leading the squadron over Negros Island, encountered his first victory which he related in his individual combat report:

"I was leading a squadron of 16 P–47s as area cover over Bacolod A/D, Negros Island. We went in at 10,000 feet stacked up to 15,000 feet. After covering the area for about 10 minutes, I took two flights down below the clouds because of the poor visibility at the altitude we were. At 5,000 feet we sighted a ZEKE which had just taken off and was heading towards the sea at an altitude of 1,000 feet. I Started down on him but was unable to get a proper lead so I didn't fire. My wingman, Lt. Pratt, however, was able to get a 60 degree deflection shot. I observed my wingman's bullets hit the cockpit of the plane which started burning and issuing a cloud of black smoke. The plane nosed over and crashed into a small wooded area southwest of Bacolod Strip on the same shore. The plane exploded upon hitting the ground.

"While making his breakaway, Lt. Pratt saw another plane pass under him and head West towards the clouds. He was shot down by another squadron in the area.*

"We proceeded along the coast and headed south, finding nothing, we returned to Bacolod approximately 10 minutes later. I was heading west toward the largest strip at an altitude of about 1,200 to 1,500 feet with two flights stacked up and acting as top cover. I saw a Type 100, 2 engine reconnaissance plane (DINAH) heading east toward the northeasternmost strip on Bacolod Drome. He was flying at about 300 feet. At 300 yards I gave him a 2 second burst and observed numerous hits in the cockpit area of the plane. At 200 yards, I gave another 2 second burst observing hits in the cockpit area. At this time he was about 50 feet over the ground over the end of the strip. He lost control of his plane and crashed by the side of the strip, nosed up twice and slid to approximately 25 yards away from the east side of the strip. I came back and strafed but was unable to start a fire. The plane was sitting on its belly and parts were scattered around the area. My wingman, Lt. Graber, shot three men, apparently crewmen who were running away from the plane. Lt. Graber observed all hits and crash of the plane. I claim one Type 100, 2 engine reconnaissance plane destroyed."

On a return engagement over Negros on November 24, Lts. Boland and Doody each claimed a victory and Lt. Ken Dunn got another Japanese aircraft on November 26. Victories were also scored by the 39th during the longest over-water flights in the squadron's history. They flew their Thunderbolts from Morotai to as far west as Tarakan, Borneo and north to the upper end of Negros Island.

When enemy bombers began harassing the bases at Morotai, a fighter sweep was arranged to catch the Japanese raiders as they were staging around Halmahera Island. One flight of the 39th took off from Wama Strip at Morotai and arrived over Halmahera late in the afternoon of Dec. 27. Just as expected, Lts. Idon Hodge and Carl Worley caught three enemy bombers over Djailolo airdrome a little before dusk. Hodge claimed two and Worly got the third. It was fitting and symbolic that two years after the first big P–38 battle in the Pacific that the squadron should cap its scoring for 1944.

Twenty-five escort missions had been flown in December in comparison to twenty patrol missions and several others of various types. It became apparent that the squadron was becoming more involved in the long range missions which would bring them in closer proximity of the enemy. The morale of the 39th was never deficient, but the prospect of renewed aerial victories caused spirits to soar.

Another factor that entered the thirst for aerial combat later in the war was the method of confirmation of enemy aircraft destroyed. If some of the squadron claims of the early days were less than absolutely confirmed then the later claims were more scientifically determined. Experienced intelligence officers and better gun-camera evidence forced the pilots to practically bring the downed aircraft back under their arms as proof. An example is the Japanese fighter claimed by Lt. Dunn over Carlotta airdrome, Negros on November 26. Dunn reported that he "kept firing even after the Tony was on fire".

---

*This aircraft was probably the Zero claimed by Capt. William H. "Wild Bill" Strand, of the 40th Fighter Squadron.

Perhaps the diligent American pilot felt that his gun-camera film should show the Japanese plane as thoroughly dispatched as a well-chewed bone.

On December 31, 1944, the ground echelon of the 39th loaded aboard LST 559 for the move to Luzon. The air group continued to operate from Morotai until it flew up to Mangaldan on January 22. At last the 39th was on Philippine soil-where all the action was. This proved a disappointment. The war in the air had just about ended for the Fifth Air Force. After some big air battles over Clark Field in December, there would never be any extensive aerial combat again in the area. In fact, for the rest of the war, the whole Command would score more than ten victories in a single day but about six times. One of those days, however, would include the 39th.

It was back to ground support for the 39th in the continuing effort to force the Japanese back into northern Luzon. During one ground attack mission on January 26, Lt. Paul Foster's P–47 had been damaged over the northeast coast of Luzon. He crash landed, then waved vigorously to his relieved comrades hovering overhead. The Thunderbolts stayed as long as possible but were forced to return home when their fuel ran low. Foster was spotted the next day on a small lake southwest of Baguey but rescue attempts were frustrated. He was sighted again sometime later and once again rescue was prevented by difficult terrain or missed coordination between Foster and the searchers. Carol Rymer searched from the air over the area where Foster's P–47 went down the next day and was surprised to find that the big fighter had been dragged away.

Hope for the downed pilot dimmed as the dangers of Japanese troops, native bandits and hostile countryside proved too much to overcome. Lt. Paul Foster was listed as missing in action,

January 30, 1945, was one of those days when the 39th helped the Fifth Air Force score in double figures. Lt. Grosshuesch was leading eight P–47s on a sweep over Formosa when enemy aircraft appeared over Taichu airdrome. Grosshuesch scored first firing at one of the biplanes, sending it spinning into the ground. The squadron had become scattered but after reforming, Grosshuesch sighted another biplane and sent it down. Then mayhem broke out. Charles Posey, flying tail-end charlie position, encountered a Japanese fighter over Taichu and fired a 90 degree deflection shot. It found the mark and the enemy fell uncontrolled to earth. He immediately engaged another fighter between Taichu and Shinchiku and sent that one down in flames. When a third fighter appeared before his guns at low altitude, Posey managed another victory for the day. The Japanese plane skidded and crashed into the trees about a mile from Shinchiku airdrome.

Gene Haws was flying Grosshuesch's number three position and managed to explode another fighter over Shinchiku. At the same time, Idon Hodge sent another fighter down in flames, Ervine Pratt apparently killed the pilot of yet another that fell into a wild spin and crashed. Haws' wingman, Eli Tueche, Jr., caused the ninth and final Japanese aircraft to crash and explode on impact. Total for the outing was two biplanes and seven fighters.

The next day, on an escort mission for a formation of B–25s, eight P–47s of the 39th ran into six Japanese interceptors about fifteen miles south of the island of Formosa. Lt. Marcus Trout quickly dispatched one fighter into the water with a four second burst from three-hundred yards away. Lt. Albert Wiget, on his first combat mission, was even more efficient, disposing of two fighters with a few short bursts.

Leroy Victor Grosshuesch as an air cadet, became last ace of the 39th F.S. and C.O. in Aug. 1945. Nat. Archives

At this late part of the war aerial victories for the Allies were few and far between but once contact was made it was not difficult to gain the upper hand. Grosshuesch's two victories, for example, were more indicative of his keen eyesight than flying skill, even though he did have a great reputation as a fighter pilot. There developed a greater competition from his own squadronmates in pursueing the enemy, for the 39th was facing a brave but painfully inadequate enemy in those final stages of the conflict.

Lt. Bolin of the 35th F.G. crashed his P-47D on Morotai, Dec. '44.

Group Commander (foreground No. 10) and Squadron Commander (rear No. 20) flank a third P-47D during an aerial portrait session.

USAF

Ranging from the Philippines, Feb.-Mar. '45

The 39th flew 231 sorties in January and 424 in February. After the squadron had settled in at Mangaldan there was more opportunity to concentrate on combat missions. Three operational days were lost due to heavy rain but the rest of the month was devoted to fighter sweeps, dive-bombing, cover for search and rescue PBY's, and local patrol.

The 39th also began ranging out to new targets closer to Japan. The top brass of the Far Eastern Air Force were eager to put the Fifth and Thirteenth Air Forces in the forefront. General Kenney was especially keen to get his aircraft ranging outward since the Seventh and Twentieth AF's were soon to mount a full-scale offensive against the Japanese home islands and the Navy was already attacking targets on Formosa. Kenney wanted his fighters and bombers in on the kill which he felt they justly deserved.

Leroy Grosshuesch led five P– 47s of the 39th on a sweep over Formosa on February 10. They bored in over the airfields of the island, spoiling for a flight. At about 3 o'clock in the afternoon, their feiste exuberance was rewarded. About twenty-five trainers were sighted over Kobi airdrome ranging from 3,000 feet down to the deck. It was a piece of cake and Grosshuesch took advantage of the situation. He led the flight down on a screaming, merciless bounce. Lt. Edgar Beam was the first to score when he fired a short burst and one of the trainers went straight into the ground. Lt. Robert Rohrs observed Beam's victory and then fired at another helpless trainer. The Japanese went straight into the ground, then burst onto flame. Lt. Charles Covey followed suit and fired at another trainer which caught fire just as he passed over it.

The party quickly broke into a complete rout. The Japanese airmen were not capable of fighting or evading the American fighters. Grosshuesch overshot one trainer only to find another directly ahead of him and dispatched it with a single burst. Covey, following Grosshuesch, noticed another trainer flying off to one side. He managed to slip easily behind and with a short snap on the trigger sent yet another trainer down in flames.

As the remaining enemy aircraft dispersed, Grosshuesch called his flight to form-up, but Lt. Winfield Langhorst noticed a Tony fighter a few miles north of Kobi. He went after it, asking Grosshuesch to cover him. The aggressive Langhorst drove right up the tail of the Tony to fire several bursts into its wings and fuselage. The Japanese fighter crashed on a river bank.

About the same time, Grosshuesch spotted a straggling trainer, easily got on its tail and sent his second victum of the day and fifth total victory spinning into the ground. Lt. Beam was also in the fray now and broke after another struggling trainer further to the north. The aircraft exploded after a short burst to become the seventh trainer in the series to be shot down. Two parachutes appeared from the disabled plane but when Lt. Rohrs went down to make a pass at them, he was horrified to see only empty canopies drifting to earth. Lts. Rohrs and Covey then attacked two more trainers but could only claim them as probables.

As the flight again reformed for the trip home, Covey spotted a Dinah, covered by a Zero, and called for action. As he went into his dive, Langhorst followed, but the Japanese spotted the pair quickly. He tried to escape but Covey's Thunderbolt caught him easily. The Zero's engine was set afire and crashed near a railroad track west of the Taichu airdrome. Covey was then able to close on the Dinah but was frustrated when he ran out of ammunition. He pulled up in exasperation and let the others have their chance. Langhorst and Beam made a pass and started one of the engines burning on the bomber. Then Lt. Rohrs had a turn only to find that he too was out of ammunition. Grosshuesch roared in and fired a few bursts sending the Dinah nosing over and exploding south of the airdrome. Low on ammunition and fuel Grosshuesch called the turkey-shoot off and headed the squadron home. Some twenty other Zeros were spotted as they left the area, but they would have to await another time.

This action would be the last major engagement the 39th would participate in during World War II. The sweep was bold; five lonely Thunderbolts had ranged freely over enemy bases, accounting for ten enemy aircraft. Had the twenty Zeros appeared earlier the story may have been different. The 2,500th Fifth Air Force victory is generally credited to Capt. Grosshuesch during this action.

Three days later another 39th formation ran into a similar situation but the aircraft encountered were the latest and deadliest in the Japanese Army air arsenal, Ki– 84 Hayate (Gale) fighters, known to the Allies as "Frank". Capt. William Rogers was leading the squadron on a cover and fighter sweep over Formosa on February 13 when he sighted nine Franks and a Sally bomber. He led the group down on the fighters and was surprised when they took no evasive action. If the Oscars he had met in November proved troublesome then these new fighters appeared to react just the opposite. Without difficulty Rogers shot one down, confirmed by Lt. Sam Teese, then pulled up with oil on his canopy. In spite of obstructed vision, Rogers was able to shoot down another Frank, then watched as Teese destroyed another with a well-aimed burst.

Lt. Marcus Trout got into the act by pulling up on the tail of a rigid-flying Frank to send it down in flames. When it was Lt. Ernest Johnson's turn he was able to send yet another Frank down into the side of a mountain. Lt. Kenn Dunn damaged another Frank badly enough to cause the pilot to bail out. Lt. Jones was flying Dunn's wing and became curious about the Japanese pilot descending in the parachute. He took his P– 47 into a wide diving arc and was shocked to witness the pilot unbuckle himself and drop from beneath the canopy, hundreds of feet from the ground. Whether he was frightened out of his senses or humiliated at his defeat, the Japanese chose not to be saved.

Six Franks were claimed by the 39th that day. One explanation for the lack of evasive action on the part of the Japanese pilots is that they may have been part of an operational training unit which was stationed on Formosa. The Sally bomber was probably tutoring as a navigational aid.

Even with the increase in aerial combat a great deal of ground attack operations were still carried out by the 39th. The Japanese were inexorably being forced back into northern Luzon and the entire 35th Fighter Group was giving support to General Krueger's First and Fourteenth Corps. The 39th participated in attacks on the mountainous terrain around the Ville Verde Trail, Balete Pass and Ipo Dam areas. An example of the support operations undertaken by the squadron is that of February 21 as described in the mission summary:

A. Mission No. 52-C-3 Squadron No. 502 39th Fighter Squadron 35th Fighter Group 21 February 1945 8 Victors.
B. Dive bomb - SE of Nichols Field and North of Bayanbayanan (14°28'N - 121°03'E)
C. Takeoff 0950/I. Landed 1125/I. Time over Target 1040/I to 1100/I. Altitude 5-1,500 feet.
D. 2 smoke bombs were put up and flight ordered to drop bombs in smoke area. Flight made run from west to east.
   7 bombs fell in target area.
   2 bombs fell just east of target.
   1 bomb released manually and over shot the target.
   1 bomb hit building in town after falling off accidentally.
   1 bomb fell on road north of town.
   1 bomb did not go off.
   3 bombs did not release over target and were jettisoned in water.
E. Thru I. NIL.
J. Manila Bay area - clear.
   over base - clear.
K. NIL.
L. 16 x 1,000 lb G.P. bombs expended. Instantaneous nose and .025 tail fuse.

The report is typical of the missions flown by the 39th during ground attack duties. In the period between January 1944 through June 1945, the 39th was engaged in these operations on the average of about fifty a month. It seems difficult to believe that more squadron personnel were not lost in view of the dangers involved. The portion of the mission summary labeled "E. Thru I. NIL." is partly concerned with enemy resistance. On this particular day no aircraft or ground fire was encountered.

Later in the month a significant air action did occur. On February 25, Capt. Grosshuesch led four P— 47s, consisting of Lts. Hodge, Johnson and Rohrs, over northern Formosa around 5 o'clock in the afternoon. Two Dinahs were sighted near Chikunan about 500 feet above them at 7,000 feet. Grosshuesch ordered the flight to drop and climb above the enemy. When the flight got just above and behind the bombers, one of them started violent evasive action while the other started to dive into the overcast. Grosshuesch and Johnson took on the first Dinah, the second was chased by Hodge and Rohrs. Grosshuesch opened fire at 300 yards with 30 degree deflection. His fire missed until the third burst which tore big pieces off the Dinah and set the right wing on fire. He followed the burning plane into the overcast and Lt. Johnson stayed right with him. The Dinah pulled back up through the clouds with Grosshuesch tenaciously chomping at his tail. He opened fire at one-hundred yards from dead astern which set the left wing on fire. One of the bomber's crew took to his parachute just as the plane turned steeply to the left and went into a dive. It crashed about thirty miles east of Chikunan. Lt. Johnson witnessed the explosion in the darkness far below.

The flight landed back at their base at about 9 P.M. The victory claimed by Grosshuesch was after dusk had fallen on the ground. It is believed that his was the first such operation involving P — 47s at night during the Formosa campaign.

March brought an increase in dive-bombing missions. Northern Luzon got its share of attention from the 39th. There were more sorties to strafe, dive-bomb and napalm stubbornly entrenched Japanese positions. Ground forces struggled for every inch of ground and the Thunderbolts of the 35th Fighter Group were involved in every grueling moment.

There was at least one happy occasion in March. About six weeks after he had been reported missing on the January 26 mission, Paul Foster came walking into camp. All bets had to be cancelled when the resourceful pilot told how he managed to avoid all the dangers of the Philippine countryside to find his way back to the American lines.

R.M. Hill

P-47 Thunderbolts of the 35th F.G. await transfer to the 58th F.G. during transition to the P-51 Mustang, March-April 1945.

OPERATIONS WITH THE P-51

There is some confusion about the exact date that the 39th traded their P–47 for P–51 Mustangs. Most sources state that the entire 35th Fighter Group started operating P–51s in March but 39th records and the memories of squadron pilots put the date at sometime in early April. The issue is not that important except that it is difficult to determine whether the 39th's sole victory in March was the first for the Mustang or the last for the Thunderbolts.

That one March air victory occured when the 39th once again found good hunting during a sweep to Formosa on the 25th. Capt. Grosshuesch was again leading the squadron when a Zero was sighted and he roared onto its tail. He managed to put a good burst into the Japanese fighter before excessive speed forced him to overshoot the target. When he looked back he saw Lt. Ervine Pratt already finishing the job and the Zero went down.

Whatever the circumstances of Pratt's victory, within two weeks brand-new P–51 Mustangs had replaced the trusty old Thunderbolts. Pilot reaction to the new aircraft was mixed at first. Old hands looked at the rugged dependability of the P–47 with its practically invulnerable air-cooled engine and massive firepower and swallowed hard at the prospect of flying long overwater missions in a lightweight Mustang with its sensitive inline Merlin engine. Even though it sparkled in air battles over Europe, the Mustang did not gain easy acceptance with the Fifth Air Force.

New pilots being trained in the United States were conditioned to the P–51 and their entry into combat was much easier. There was also the anticipation of an invasion on the Japanese homeland and the Mustang was considered more flexible in adapting to these demands.

Climax in the Philippines, Apr.-June '45

The younger pilots did not entirely dominate the image of the squadron in 1945. Some of the old hands began to reappear during the early spring. Stanley Andrews came back briefly while he prepared to take command of a new squadron for the planned invasion of Japan. The 39th was also unique in having two sets of brothers in its ranks. George and Gordon Prentice had flown together briefly before George went on to command the 475th Fighter Group. During 1944 and into 1945, James and Robert Querns flew together during squadron operations. Late in July 1945, Leroy Grosshuesch relinquished command to Maj. Ben Widmann, another veteran of the 39th. Grossuesch remained with the squadron however, and became C.O. again after the war.

It was generally believed that Japan would not capitulate short of a crushing Allied assault on the home island. The country was all but cut off from the necessary resources to carry on a war and was slowly strangling to death. Her merchant fleets no longer existed and the Allied blockade was tightening barely a few hundred miles from her shores.

Early in April, the 39th moved to the Lingayen area. By the twenty-first, the squadron had moved with the rest of the 35th Fighter Group to Clark Field outside of Manila. Throughout May, strafing and bombing attacks were carried out against Japanese strong points in northern Luzon. By the beginning of June, however, a major breakthrough was made in the Cagayan Valley and the enemy's defenses crumbled. Aside from routine mopping-operations, Luzon was secured.

The entire 35th Fighter Group was pulled out of action following the fall of the Philippines and throughout the remaining month of June prepared to transfer to a new base. By June 27 the 39th was on its way to Ie Shima, just off the coast of Okinawa. They were operating from Japanese real estate. It would be the Group's final station in World War II.

Combat over Japan, July-Aug. '45

General Kenney, itching to get his fighters into combat over the Japanese mainland, lost no time in assigning operations which would permit contact between Fifth Air Force fighters and the defenders of Japan. Although it was possible that the Japanese would horde their remaining airpower for the impending invasion, Kenney believed that there would be plenty of aerial action before that came about.

Dick Cella had returned to the Pacific fighting and became the 35th Fighter Group's operations officer. He led the first group of forty-eight Mustangs over Kyushu on July 3. Contact was made almost immediately with enemy fighters. Flying with his old outfit, Maj. Cella downed one of the defenders. Lt. Winfield Langhorst shot down another and Capt. Bob Mittelstadt claimed two others to give the Fifth Air Force its first four victories over Japan, courtesy of the 39th Fighter Squadron.

New P-51D Mustangs for the 39th F.S.
First aircraft; pilot Lt. Charles Posey
2nd; "Barbara Jean," Lt. W.F. Cook
3rd; "Miss Allayne-ious," Lt. W.S. Langhorst
4th; "Badger Beauty,"
5th; "Mannequin Joanne," Lt. Howard Neuman.
Squadron color was medium blue.
Most aircraft names were medium blue, thinly outlined in red.

USAF

"Miss Sandra," Capt. Ernest E. Johnson "Screamin Demon," 1st Lt. Ervine C. Pratt. Black wing and fuselage bands were not outlined in another color, removal of masking tape left a false one inch border.

F.F. Smith

General Kenney was so ecstatic over this news that he called for a parade on July 7. The scrappy, gravel-voiced little general could hardly disguise his pleasure when he decorated Cella, Langhorst, Mittlestadt and others who had participated in the mission. Whatever their expectations though, the 39th would never fight a pitched air battle again in World War II. Aerial victories at this late point in the war were extremely rare for all Allied Air Forces. Another victory was claimed by the 39th on July 15 during a submarine cover mission. Capt. Mittelstadt was in the forefront and his mission report reads:

A. Mission No. 196-C-5-A. Squadron No. 845. 15 July 1945. 39th Fighter Squadron, 35th Fighter Group APO 331. 8 P - 51's.
B. Cover Sub at 30 miles SE of Reference Point K-3.
C. T.O 1005/I. Landed 1505/I. T.O.T. 1115/I to 1215/I.
D. After covering sub, flight went on a Fighter Sweep over Southern Kyushu. Definitely destroyed - one single-seated fighter (George 11) 2 miles NE of Kumatto (32°45'N - 130°43'E) A/D at approximately 1300/I.
E. Interception - Yes. 1 type single engine fighter (George 11) at 1300/I, altitude 300 feet, 2 miles NE of Kumamoto A/D.
F. One bogey was called in to the squadron leader at 6 o'clock low. Captain Mittelstadt sighted the bogie and started down from 10,000 feet. The plane was identified as enemy immediately for it was taking off from the strip when first sighted. When Captain Mittelstadt was 350 yards from the enemy plane, he opened fire. His position was above and to the rear of the George. Captain Mittelstadt observed pieces flying off the enemy plane and saw it burst into flames. The George rolled over to the right, crashed into the ground and exploded. This was approximately 2 miles NE of the A/D. Lieutenant Lemke observed the action.
G. About 4 bursts of medium A/A from Nagasaki (32° 1'N-129°55'E) accurate as to altitude.
H. NIL.

Two weeks later Grosshuesch and Widmann were on a similar fighter sweep. The summary for 35th Fighter Group mission 211-C-3-A, dated 30 July 1945 indicates that the two pilots, for want of air targets, went down to strafe a pair of Japanese destroyers off the coast of Japan. A few days later Leroy Grosshuesch was back over the area and capped the victories tally for the 39th Fighter Squadron and the 35th Fighter Group. Flying over Kyushu, he sighted a Frank and managed to get on the Japanese fighter's tail but spun out in the excitement of another kill. He managed to pull out and get back into position. A moment later he sent the Frank down to destruction. The date was August 12, 1945, just three days before the war was declared over.

A footnote to Grosshuesch's last victory is that some of these claims, made after August 1, are not included on most Ace's lists of the period. Grosshuesch had been listed with seven victories when in fact he has eight confirmed credits. Albeit even the most confirmed air victories are arbitrary at best and many claims are now doubtful what with a comparison of records becoming available from the other side.

39th F.S. Postscript, Autumn '45

By October 10, 1945, the 39th had moved along with the 35th Fighter Group to Irumagawa, Japan and became part of the occupation force. An accident occured during this time that sadly marred an otherwise jubilent squadron. Paul Foster had decided to stay on with the squadron after returning from his trek through the Philippine jungle. His luck failed him, however, when his P— 51 malfunctioned and he went down into Tokyo bay in a fatal crash.

In spite of the loss of many such fine pilots, the 39th could boast of an unflagging service record. It was one of the first American fighter squadrons to enter combat in WW II, fighting valiantly with the P — 39 Airacobra. They introduced the P — 38 Lightning to the New Guinea campaign with devastating results, hammered the enemy relentlessly with the P — 47 Thunderbolt, then carried the battle to the enemie's doorstep with the P— 51 Mustang. But the real glory of the 39th rests in the spirit of its members and their willingness to take the fight to the enemy under any circumstances. Their ability to accept and make the most of a variety of assignments and equipment attests to the 39th's proud contrubution to the war effort.

World War II Postscript, 1946-53

The first winter spent in Japan was cold and miserable. Most of the crews were quartered in wooden huts, heated with dangerous kerosene stoves that accounted for several disasterous fires. It was typical postwar service; once the emergency was over, funds were cut, publicity ended and overseas outpost duty had to put up with low priorities in comfort and promotions.

But the duty was not bad. Without the urgency of war there was time to become aquainted with Japanese culture and the 39th had its pick of brand-new P–51D aircraft. The squadron performed well in postwar parades and air demonstrations, and kept abreast of the latest combat tactics.

Leroy Grosshuesch again assumed command of the squadron at the end of 1945. He stayed at the helm until the spring of 1948 when he returned to the U.S. and eventually retired from the Air Force. Grosshuesch had won his wings on July 28, 1943, and by the end of 1948 had spent most of his illustrious career with the 39th.

Other postwar commanders of the 39th were Ervin Ethell who stayed in the service and retired as a colonel, and Gerald Brown who commanded the 39th through its first tours in the Korean war. Major Brown was shot down in his Mustang and endured captivity by the North Koreans until repatriated after cessation of hostilities.

J. Ethell

Proud entrance sign at Irumagawa, Japan 1949.

J. Ethell

-51Ds of the 39th F.S., part of the ccupation forces based at Irumagawa, apan. Above; markings right after rmestice and right, the flight line in 949.

The Korean conflict

Participation in the Korean war began quickly for the 39th because of its location in nearby Japan. Within a few days, however, the squadron was forced to bounce from its base in Pohang back to Japan in the face of the speedy North Korean advance. They returned to Korea in October after the Inchon landing and stayed around for some final glorious moments until the end of the conflict.

One of the things that the pilots did not particularly like about the first tour was the F– 80C jet fighter assigned to them. Although it was fast and carried an appreciable load of weapons, the Shooting Star had a short range and needed long smooth runways to get airborne. The 39th pilots were used to the piston engined Mustangs that had equipped the squadron for years. They wanted them back and finally got their wish, flying missions throughout 1951 in old P-51s.

J. Ethell

Lt. Col. Gerald Brown, C.O. when the 39th was again called to duty in Korea, displays latest hard hat flight gear.

Once again the 39th was called upon to carry out ground-attack missions. From the beginning of the war until they stood down in the latter days of 1951, the 39th flew a hectic series of combat sorties much like their days with the P– 47 in World War II. Early in 1952 the 39th was reequipped with F — 86F fighters, the fighter-bomber version of the Sabre jet. By June they were transferred from the 18th Fighter Bomber Wing to the 51st Fighter-Interceptor Wing. It was a most timely conversion for the opposition north of the Yalu river were just starting a new air offensive with an agressive crew of Mig fighter pilots. September was the high point of the offensive which saw the loss of no fewer than thirteen U.S. jets for a record bag of sixty-three Migs.

The Migs became more cautious during the next few weeks but the 39th had scored their first victory. After November the action picked up for the 51st Fighter-Interceptor Wing which saw its first aces created in December. During November and December at least fifty-six Migs were claimed, some of which were clearly marked as soviet aircraft.

Late in January 1953 the 39th produced its first fighter ace of the Korean war when Captain Harold Fischer downed his fifth Mig. He downed two more on February 15, during an escort of F – 84 fighter-bombers. On the mission, Fischer sighted a number of Migs and was startled to find himself flying abreast of one. Sabre and Mig immediately began scissoring maneuvers without either finding an advantage for several minutes until Fischer took a risk and popped his dive brakes which slowed the F– 86. He fell in behind his foe and fired until the communist pilot bailed out.

Capt. Harold E. Fischer became 25th jet ace in the Korean conflict; had 10 enemy MiGs to his credit. The Sharkmouth carry-over from WW II was a personal decoration. J. Lane

Capt. Fischer then spotted another Mig on the tail of an F– 86 and gave chase. The Mig pilot reacted quickly and ran for the Yalu. When Fischer closed the distance the Mig tried a zooming turn. The effort was not enough as Fischer fired a few telling bursts. Both American and Communist jets went into a stall spin but Fischer recovered while the Mig went in near the Chongchon river.

On the next day, 1st Lt. Joe McConnell, Jr. downed his fifth Mig. McConnell had always wanted to be a fighter pilot but was frustrated during World War II in his cadet days when he was assigned navigator training and eventually flew a tour over Europe in B–24s. When the new U.S. Air Force began its expansion program, McConnell seized the opportunity for pilot training. In 1952 he was sent to Korea as an F – 86 fighter pilot and joined the 39th when it re-entered the war as a Sabre-jet outfit.

Action in the Korean skies remained heavy during March although the Migs were beginning to show reluctance to enter combat. For the most part the Sabres escorted B — 29s or F – 84s on bombing missions. Later in the month, Harold Fischer was credited with his ninth Mig and McConnell got his seventh.

April was a slow month; still twenty-seven Migs decorated the Korean countryside – five courtesy of the 39th. However, both Fischer and McConnell were numbered among the Sabre pilots downed during the month. McConnell had just claimed his eighth Mig when he was forced to glide his battle-damaged F– 86 over the Yellow Sea and eject. A helicopter rescued him soon after. Fischer was not so lucky. His F – 86 was hot after its final victory north of the Yalu when it was hit by cannon fire and he was forced to parachute into enemy hands. Fischer was regarded as a criminal for being caught in Manchuria and held prisoner until 1955.

First jet aircraft used by the 35th FIW (39th FIS) was the Lockheed P-80 Shooting Star. A pair from the 35th Wing are seen here leaving their base in Japan for a rocket attack on North Korean supply lines, Aug. 6, 1950.

North American F-86F Sabre. . .the type used nost successfully by American forces during the Korean conflict. Inter-service exchange pilots were common during this time making it difficult to pin down pilot and aircraft assigned to any particular squadron at a given time period.

"Mac" McConnell continued to add Migs to his tally in spite of the dunking he received. He claimed two more during April to give him ten. On May 13 he added his eleventh then got a twelfth victory two days later. The very next day "lucky 13" fell to McConnell's guns.

Sabres of the 4th and 51st FIW were patrolling over the Yalu with a vengeance during the spring of 1953. May 18 was a red-letter day for the 39th when Lt. Col. Ruddell claimed his fifth Mig to become the third ace of the Korean War. At about the same time, McConnell had his flight at its patrol height of 46,000 feet over the Yalu when ground control called to inform them that about thirty Migs were in the area. The squadron located four as they crossed their path. Mac latched on to the number two Mig. In a twisting and turning dogfight McConnell finally forced the Mig into a dive and fired at the aircraft until the pilot jettisoned his canopy and ejected. In the meantime, the Mig flight leader slipped in behind Mac and opened fire with everything he had. It was a real dogfight until the Sabre pilot was able to outmaneuver his adversary. As he closed in from behind, the Mig shuttered as bullets ripped into the engine. As the Mig started smoking, Mac managed to get in close enough to see the pilot's face. The desperate enemy then tried to ram McConnell before he spun in and crashed. With that McConnell was the first American pilot of the Korean war to score fifteen confirmed victories.

During the afternoon another sweep was flown over the Yalu and Mac scored his sixteenth Mig and was immediately grounded to be hailed as the top American jet ace. By the end of the war in Korea, four of the top five aces of 51st Fighter Interceptor Wing would be 39th pilots. James Jabara of the 4th FIW was the only pilot to seriously threaten Mac's record in subsequent combat and is the only other American triple jet ace.

June and July produced the last aces of the 39th in Korea. Lt. Henry Buttleman destroyed his fifth Mig in June and TDY (temporary duty) Marine Major John Bolt became an ace in July. Bolt was already an ace with Boyington's "Blacksheep" VMF 214 during World War II when he was assigned to the 39th. He was an aggressive and daring fighter pilot who was reported to risk court martial at times to get at the enemy. He flew Navy F9F Panther jets before the opportunity came that enabled him to be assigned to McConnell's "hot" Dog flight.

On July 11, 1953, Bolt was leading Dog flight on a cover mission for a reconnaissance over Mig bases north of the Yalu. No Migs had been sighted by the 39th for ten days but Bolt rejoiced when he sighted four interceptors coming up to meet them. Bolt and his wingman dropped their external tanks and rolled over to take on the enemy. The second Mig on the left became Bolt's target and went down smoking to crash after a few bursts of .50 calibre fire. Meanwhile, Bolt's wingman had attacked the Mig leader and forced him into a favorable position for Bolt to make a pass. The Sabre-flying marine closed to within one-hundred yards and fired directly up the jet's tailpipe until it smoked violently and the pilot ejected to float downward in his parachute. With their fuel at the critical point, the two Sabre pilots headed home to Kimpo.

Within three weeks the Korean war was over. Never again would the determination, skill and courage that had been the tradition of the 39th be required in armed combat. At present, (1982) the 39th is still active as the 39th Tactical Fighter Squadron, prepared for whatever the future holds but hoping their combat history is over.

It is possible to reflect on the technical expertise of Dick Cella or Ben Widmann; or the geometrically logical mind of Tommy Lynch, the little pugilist from Pennsylvania, who instantly won the respect of the young men he led while he won victory after victory in the air. Then there were the other leaders, such as George Prentice who directed the 39th through its first period with the P–38, or the soft-spoken and mild Charles King who held a tight rein on the exhuberance of wild-flying eager pilots and kept the squadron operating as a smooth team. Also Leroy Grosshuesch, who led determined sweeps to find the diminishing Japanese aerial presence and claimed some of the last air victories of WWII.

There was Curran "Jack" Jones, who tackled an ace Japanese Zero pilot without hesitation for his first victory. Ken Sparks was an eager wildcat who probably brought home more damaged P–38s than any other 39th pilot because of his manner of throwing himself into a dogfight without any visible fear. He finally drew an unfortunate lot when he tried a risky dive over the coast of California in 1944 and crashed to his death in the Pacific Ocean. The two top aces of World War II and Korea scored with the 39th and also met their deaths in subsequent accidents. Dick Bong lost his life testing a P–80 in 1945 and Joe McConnell died testing an F–86 in 1954.

39th F.S. pilots, summer 1943.
L. to R.; R.E. Smith
unknown
unknown
unknown
P. Stanch
K. Sparks
G. Prentice
J. Lane
A. Kish
last two unknown

L. to R.; Eli Tueche, Jim Lockhart, Lt. Doady, Charles Covey and Paul Larrick.

39th billboard on Morotai, Dec. 1944.

Leroy Grosshuesch became last WW II 39th ace in this P-51. Note dark blue command stripes and spray marks left after masking tape was removed. This masking effect was sometimes interpreted as a colored border. Below: Returning from a fighter cover over Japan P-51 Mustangs of the 35th Fighter Group peel off for a landing at their home base, Ryukyu Retto, Okinawa, July 1945. Navy PB4Y-2 patrol bombers line the runway.

In all, the 39th garnered four Distinguished Unit Citations in World War II and Korea; one for Papua in July 1942 and another for the Battle of Bismarck Sea plus two for the various tours in Korea. There were also two Republic of Korea Presidential Unit Citations and another from the Philippines. Throughout its combat career the 39th gained many enviable credits. Bob Faurot is perhaps the finest example of the 39th's spirit. He fought off both jungle disease and the fatigue of an arduous operational tour while encouraging other members of the squadron to new limits of endurance. His death came about because he led his flight into a superior force of Japanese fighters for the benefit of a B–17 in his charge. The 39th Fighter Squadron has a long history and a proud heritage indeed.

Lt. Paul Foster shortly before he was downed over Luzon in January 1945. While he survived the crash and trek back to civilization, he lost his life when he went down with his P-51 over Tokyo Bay after the war.

Bob Faurot, who dropped a bomb off Lae airstrip that exploded under a Zero taking off. He died in fighting off a swarm of enemy fighters while valiantly protecting a lone B-17 under his protection.

Pilots of the 39th gather for the usual bull-session during the P-47 era.

WW II ACES, WHILE ASSIGNED TO THE 39th FIGHTER SQUADRON

Thomas Lynch	16
Kenneth Sparks	11
Paul Stanch	10
Leroy Grosshuesch	8
Richard Smith	7
Stanley Andrews	6
Hoyt Eason	6
Charles Gallup	6
John Lane	6
Charles King	5
Curran Jones	5
Richard Suehr	5
Charles Sullivan	5

THOMAS J. LYNCH *

20	May	42	2	Zero
26	May	42	1	Zero
27	Dec	42	2	Oscar
31	Dec	42	2	Zero
8	Jan	43	1	Oscar
3	Mar	43	1	Zero
8	May	43	1	Zero
10	Jun	43	1	Dinah
20	Aug	43	2	Nick
21	Aug	43	1	Oscar
4	Sep	43	1	Dive Bomber
16	Sep	43	1	Dinah

KENNETH SPARKS

27	Dec	42	1	Zero
			1	Val
31	Dec	42	2	Zero
7	Jan	43	1	Oscar
8	Jan	43	2	Oscar
4	Mar	43	2	Fighter
18	Jul	43	1	Oscar
21	Jul	43	1	Oscar

PAUL STANCH

3	Mar	43	2	Fighter
4	Mar	43	1	Fighter
21	Jul	43	2	Oscar
23	Jul	43	1	Zero
2	Sep	43	1	Twin-engine
22	Sep	43	2	Oscar
23	Oct	43	1	Zero

RICHARD E. SMITH

6	Jan	43	1	Oscar
3	Mar	43	1	Zero
12	Apr	43	1	Betty
21	Jul	43	2	Tony
23	Jul	43	1	Oscar
22	Sep	43	1	Zero

STANLEY O. ANDREWS

27	Dec	42	1	Zero
6	Jan	43	1	Oscar
8	Jan	43	1	Oscar
3	Mar	43	1	Oscar
21	Jul	43	1	Tony
20	Aug	43	1	Zero

HOYT EASON

27	Dec	42	2	Zero
31	Dec	42	3	Zero
8	Jan	43	1	Oscar

CHARLES S. GALLUP

27	Dec	42	1	Zero
7	Jan	43	2	Oscar
8	Jan	43	2	Oscar
12	Apr	43	1	Betty

JOHN H. LANE

31	Dec	42	1	Zero
7	Jan	43	1	Oscar
8	Jan	43	1	Oscar
3	Mar	43	1	Fighter
18	Jul	43	1	Oscar
29	Aug	43	1	Oscar

CHARLES P. SULLIVAN

17	Jun	42	1	Betty
6	Jan	43	1	Oscar
3	Mar	43	1	Fighter
12	Apr	43	1	Betty
21	Jul	43	1	Nell
26	Jul	43	1	Oscar

RICHARD I. BONG **

27	Dec	42	1	Val
			1	Zero
7	Jan	43	2	Oscar
8	Jan	43	1	Oscar

CURRAN L. JONES

9	Jun	42	1	Zero
6	Jan	43	2	Oscar
3	Mar	43	1	Zero
			1	Fighter

CHARLES W. KING

2	Mar	43	1	Oscar
21	Jul	43	1	Oscar
22	Jul	43	1	Oscar
29	Oct	43	2	Oscar

RICHARD C. SUEHR

9	Jun	42	1	Zero
6	Jan	43	1	Oscar
8	Jan	43	2	Oscar
12	Apr	43	1	Betty

LeROY V. GROSSHUESCH

21	Nov	44	1	Dinah
30	Jan	45	2	Biplane
10	Feb	45	2	Biplane
			1	Dinah
25	Feb	45	1	Dinah
12	Aug	45	1	Frank

***Lynch was sent home in Sept. '43 and returned to the SWPA in Jan. '44 after which he obtained his last four victories with HQ., Fifth Fighter Command.**

10	**Feb**	**44**	**1**	**Lily**
3	**Mar**	**44**	**1**	**Oscar**
			1	**Tony**
5	**Mar**	**44**	**1**	**Oscar (Total victories 20)**

****While attached to the 39th F.S. He went on to acquire a total of 40 victories with the 49th and 475th Fighter Groups.**

THOMAS J. LYNCH
16 Victories
C.O. March 1943

KENNETH SPARKS
11 Victories

PAUL STANCH
10 Victories

RICHARD SMITH
7 Victories

STANLEY ANDREWS
6 Victories

HOYT EASON
6 Victories

CHARLES GALLUP
6 Victories

JOHN LANE
6 Victories

CHARLES SULLIVAN
5 Victories

CURRAN JONES
5 Victories

CHARLES KING
5 Victories
C.O. Sept. 1943

RICHARD BONG
5 Victories

RICHARD C. SUEHR
5 Victories

LEROY GROSSHUESCH
8 Victories
C.O. Aug. 1945

DICK CELLA
C.O. June 1944

GEORGE PRENTICE
C.O. 1942

KOREAN WAR ACES, WHILE ASSIGNED TO THE 39th FIGHTER SQUADRON

Capt. HAROLD E. FISCHER
10 Victories

1st. Lt. JOSEPH McCONNELL, JR.
16 Victories

Date	Victory
Jan 14, 53	1 MiG
Jan 21, 53	1 MiG
Jan 30, 53	1 MiG
Jan 31, 53	1 MiG
Feb 16, 53	1 MiG
Mar 8, 53	1 MiG
Mar 14, 53	1 MiG
Apr 12, 53	1 MiG
Apr 16, 53	1 MiG
Apr 24, 53	1 MiG
May 13, 53	1 MiG
May 15, 53	1 MiG
May 16, 53	1 MiG
May 18, 53	3 MiG

Date	Victory
May 5, 53	1 MiG
Jun 22, 53	1 MiG
Jun 24, 53	1 MiG
Jun 30, 53	1 MiG
Jul 11, 53	2 MiG

Maj. JOHN BOLT, USMC
(exchange pilot)
6 Victories

Lt. Col. GEORGE I. RUDDELL
8 Victories
C.O. 39th F.I. Sqdn.
51st F.I. Wing

ROSTER: ORIGINAL 39th F.S. PILOTS. . .WORLD WAR II

39th Pursuit Squadron Pilots–
July 1941

McNickle, Marvin
Faurot, Robert
Green, Donald
Royal, Francis
Lynch, Thomas
Robertson, John
Fleming, Thomas
LeBreche, George
Davis, Patrick
King, Charles
White, Edward
Tovera, Edward
Carey, Ralph
Jones, Curran
Cooper, James
Coward, James
Fink, Ord
Holloway, Authur
Little, Edwin
Mitchell, Merlin
George, Kenneth
Williams, George
Whisonant, William
Zimlich, Louis
Macomber, Louis
Vinson, Arnold

Joined Dec. 1941

Angier, Frank
Eason, Hoyt A.
Bartlett, George H.

Overseas Orders 18 Jan 1942
S.O. 18 Shipment 4580 D-39

Baluss, John W.
Royal, Frank R. (Sq. Cdr)
Green, Donald J.
Faurot, Robert L.
Lynch, Thomas J.
Carey, Ralph C.
King, Charles W.
Hess, Edward W.
Eason, Hoyt A.
Bartlett, George E.
Angier, Frank E.
Also on shipment:
Jones, Curran L.

Squadron Pilots as
of May 1942

Berry, Jack W. (Sq. Cdr)
Wahl, Eugene
Adkins, Frank
Lund, Harold
Parker, George A.
Rauch, Carl T.
Suehr, Richard C.
Sullivan, Charles P.
Hinson, Robert J.
Hylton, John
McMahon, Robert F.
Rehrer, Harvey E.
Price, John C.
Gallup, Charles S.
Ralston, Wilson
Beane, Walter O.
David, Stephen L.
Ryan, James

Jones, Curran L.
Carey, Ralph
Marlett, Wilmott
Bartlett, George
Douglas,
Shoup,
Royal, Francis R.
King, Charles W.
Lynch, Thomas J.

Joined late 1942 - early 1943

Mangas, John H.
Dunbar, John C.
Larson, Paul J.
Prentice, George (Sq. Cdr)
Kish, Andrew K.
Shipley, Lloyd P.
Randall, Edward W.
Tansing, Richard M.
Haigler, Lee
Baker, Robert H.
Harvey, Clyde L.
Stanch, Paul M.
Denton, Harris L.
Bills, Ralph C.
Flood, Edward C.
Widmann, Benjamin
Cella, Richard
Sparks, Kenneth C.
Andrews, Stanley O.
Lane, John H.
Rogers, John J.
Walters, James D.
Turick, Henry R.
Smith, Richard E.

Non-Rated

McGuire, Eugene I.
Harvey, Robert P.
Rosenburg, Raymond H.
Runser, Albert W.
Stubbs, Claude M.
Brake, Harry R.
McGettigan,
Hellmuth, John N.
Hansleman, George
Dake, Donald A.
Berry, Jack G.
Vogel, Harry

23 March 1943 S.O. 16

Hollabaugh, Simpson
Prentice, Gordon B.
Jealke, William G.
Carr, Walter W.
Dorsey, Harry N.
Steele, James F.
Vining, Leland P.
Round, Howard N.
Laing, Hamilton C.
Salmon, Hamilton H.
Weston, Raymond
Maxwell, Edward B.
Lockhart, Lewis
Mettler, William M.
Urquhart, William L.
Sander, William B.
Rothgeb, Wayne P.
Kramme, Raymond W.
Younkman, Marshall A.
Duncan, Gene
Forest, Joseph E.

39th COMMANDERS 1940-1980

Rank	Name	Date
Capt.	Allen Springer	1 Feb 1940
Capt.	W. T. Clingerman Jr	16 Jan 1941
1st Lt.	Marvin McNickle	5 Mar 1941
Lt.	Francis R. Royal	19 Jan 1942
Lt.	Donald J. Green	25 Mar 1942
Maj.	Jack W. Berry	4 Apr 1942
Lt.	Stephen David (acting)	4 Aug 1942
Lt.	Francis Royal	17 Aug 1942
Maj.	George Prentice	18 Sep 1942
Maj.	Thomas J. Lynch	24 Mar 1943
Maj.	Charles King	20 Sep 1943
Maj.	Harris Denton	18 Dec 1943
Maj.	Richard Cella	20 Jun 1944
Capt.	Leroy V. Grosshuesch	7 Nov 1944
Maj.	Ben Wedmann	Jul 1945
Capt.	L.V. Grosshuesch	Aug 1945;
		31 Dec 1945 - 30 Apr 1948
Maj.	Thomas D. Robertson	30 Apr 1948
Capt.	Robert D. Welden	Nov 1948
Capt.	Ervin C. Ethell	Apr 1949
Maj.	Gerald Brown	21 Nov 1949
Maj.	Charles Bowers	31 Jul 1950
Maj.	T.D. Robertson	30 Nov 1950
L Col.	Gerald Brown (acting)	30 Dec 1950
Maj.	Charles Bowers (acting)	28 Feb 1951
L Col.	Leland Molland (acting)	Mar 1951
L Col.	T.D. Robertson	30 Apr 1951
Maj.	M.H. Davis	10 Jun 1951
Maj.	Jack Davis	14 Aug 1951
L Col.	Francis Bodine (acting)	31 Oct 1951
Maj.	Jack Davis	Nov 1951
L Col	G.L. Gilliland	31 Jan 1952
Maj.	William H. Wescott	1 Jun 1952
Maj.	Francis D. Hessey	23 Aug 1952
Col.	George J. Ruddell	6 Nov 1952;
		June 1954 - Dec 1957
L Col.	Norman W. Campion	8 Dec 1957
L Col.	Delbert C. Hainley	16 Oct 1969
L Col.	Alvin B. Richards	9 Feb 1970
L Col.	Albert V. Van Aman Jr	12 Mar 1972
L Col.	Billie R. Oney (acting)	1 Mar 1973
		-15 Mar 1974
L Col.	James H. Martin	1 Jul 1977
L Col.	Wyrewood A. Gowell	15 Dec 1978
		-30 Sep 1980

39th Fighter Squadron

CHRONOLOGY

Feb. 1, 1940	39th Pursuit Squadron activated at Selfridge Field, Mich.
Dec. 7, 1941	Squadron moved from Baer Field, Indiana to Paine Field, Washington for overseas assignment.
Jan. 31, 1942	Squadron sails for Australia.
Feb. 25, 1942	Squadron arrives at Brisbane.
Jun. 2, 1942	Began first tour of duty at Port Moresby.
Jul. 30, 1942	First tour of duty ends, squadron rested and reequipped.
Oct. 2, 1942	Two flights of 39th P-38s escort the B-17 bringing General MacArthur to Port Moresby.
Nov. 6, 1942	Squadron officially begins operations with P-38s.
Dec. 27, 1942	Eleven Japanese aircraft claimed in first big fight using P-38 fighters.
Dec. 31, 1942	Nine Zeros downed in air battle over Lae.
Jan. 6-8, 1943	Twenty-nine Japanese fighters downed in battles over enemy convoys near Lae.
Mar. 2-4, 1943	Participates in the Bismarck Sea battle; claim seventeen Japanese fighters.
Apr. 12, 1943	Five Japanese raiders claimed in the defense of Port Moresby.
Jul. 21, 1943	39th claims twelve Japanese aircraft to become the first V Fighter Command squadron to down one-hundred enemy aircraft in aerial combat.
Sep. 22, 1943	Seven enemy fighters claimed.
Oct. 12 - Nov. 7, 1943	Participates in the Rabaul operations, claim nine Japanese aircraft.
Nov. - Dec. 1943	Reequipped with P-47 aircraft.
Dec. - Jan. 1944	Supported Sixth Army and moved to Gusap.
Feb. - May 1944	Continued ground support and moved to Lae on May 25.
Jun. 9, 1944	Returned to Nadzab.
Aug. 7, 1944	Moved to Noemfoor.
Oct. 23, 1944	Moved to Moratai.
Nov. 6, 1944	Three enemy aircraft claimed, the first confirmed victories in over a year.
Jan. 22, 1945	Moved to the Philippines.
Jan. 30-31, 1945	Twelve enemy aircraft claimed in two big air battles.
Feb. 10, 1945	Last big air battle of WW II for 39th nets ten more air victories.
Mar. - Apr. 1945	Reequipped with P-51 aircraft.
Jun. 30, 1945	Moved to Okinawa.
Jul. 3, 1945	Participated in the first Fifth Air Force sweep over Japan.
Aug. 10, 1945	Capt. LeRoy V. Grosshuesch scores last 39th aerial victory of WW II.
Oct. 10, 1945	Moved to Irumagawa, Japan.
Aug. 7, 1950	Moved to Korea.
Jun. 1, 1952	Began operations with F-86 aircraft.
May 18, 1953	Joe McConnell becomes top air ace of the Korean war.
Jul. 20, 1954	Moved back to Japan.
Dec. 8, 1957	39th deactivated at Komaki, Japan.
Aug. 18, 1969	Redesignated 39th Tactical Reconnaissance Training Squadron.
Oct. 15, 1969	Squadron reactivated.

Feb. 15, 1970	Redesignated 39th Tactical Electronics Warfare Training Squadron.
Mar. 15, 1974	Squadron deactivated again.
Jun. 1, 1977	Redesignated 39th Tactical Fighter Training Squadron.
Jul. 1, 1977	Reactivated, assigned to 35th Tactical Fighter Wing. George AFB, CA.
Dec. 1977	Last F-105Gs transferred to other 35TFW units and 39th becomes a solely F-4 unit (later F-4Gs)
Jul. 24 - Sep. 20, 1978	Period that first Wild Weascl class (F-4C) began. Final class completed training.
Apr. 2 - Jun. 25, 1979	PACAF initial Advanced F-4G Wild Weasal crews completed training.
Oct. 9, 1980	39TFTS redesignated 39th Tactical Fighter Squadron, authorized F-4Es but remained without personnel or equipment until present transfer of aircraft from 21TFW, Elmendorf AFB, Alaska.

Note: Prior to redesignation, the 39th TFTS garnered these honors and awards:

7 Feb. 1980	39th TFTS wins Turkey Shoot. 39th AMU top maintenance unit.
15 Feb. 1980	Lt. Gen. Nelson Presents Air Force Outstanding Unit Award to 39th TFTS.
15 May 1980	39th AMU scored 100 percent in maintenance events of Turkey Shoot.
11 Sep. 1980	39th TFTS wins second Turkey Shoot.

AIRCRAFT USED BY THE 39TH FIGHTER SQDN.

P-35 and probably P-36	February 1940 - spring 1941
P-39	spring 1941 - August 1942
P-38	late summer 1942 - November 1943
P-47	December 1943 - March 1945
P-51	April 1945 - spring 1950
F-80	spring - summer 1950
F-51	late 1950 - spring 1952
F-86	1952 - 1954
F-94	1954
F-86	1954 - 1957
RB-66, WB-66	1969 - 1970
EB-66	1970 - 1974
F-105G	1977
F-4C	1977 - 1978
F-4G	1978 - 1980
F-4E	1978 - 1980

Notes, memoirs, autographs